U.S. MILITARY
WORKING DOG
TRAINING
HANDBOOK

★

DEPARTMENT OF DEFENSE

LYONS PRESS
Guilford, Connecticut
An imprint of Globe Pequot Press

CONTENTS

CHAPTER 1

VETERINARY TRAINING PRIORITIES

MOTIVATION

Today's dog handlers are deployed more than ever to locations that may not always have veterinary support. This chapter is a refresher to training that you are required to receive from your home station veterinary personnel annually. If you have not been properly trained, this guidance is not the starting point for learning these skill sets and you should seek out the training. You must have prior knowledge and training of these skills or some of these lifesaving steps may do more harm than good by an untrained handler.

Take Vital Signs

Vital signs are a key component of the physical evaluation of a military working dog (MWD). As a dog handler, it is important that you be familiar with how to take your dog's vital signs. You must also know what is considered "normal" for your dog. Learning how to take the vital signs will allow you to quickly recognize abnormal conditions and relay important findings to veterinary staff.

Measure the vital signs of the dog

Vital signs are most representative of the dog's health if measured while the dog is at rest and not stressed. The core vital signs should be measured at every physical examination or when evaluating a dog because of illness or injury and should include body temperature, pulse rate and character, respiratory rate and character, mucous membrane color, capillary refill time (CRT), skin elasticity, level of consciousness, body weight, and body condition score (BCS).

- Determine the dog's body temperature using the rectal temperature measurement method.
- Lubricate the thermometer by squeezing a small amount of sterile lubricant onto a gauze sponge and rolling the thermometer tip in the lubricant.
- Lift the tail gently and insert the thermometer 1 to 2 inches into the dog's rectum.
- Support the abdomen and do not allow the dog to sit.
- Hold the thermometer in place until it beeps or flashes.
- Remove the thermometer and wipe it with a gauze sponge soaked with alcohol.
- Read the thermometer. The normal rectal temperature of a dog is 100.5°F to 102.5°F. The dog's temperature may be increased due to high environmental temperatures, stress, or exercise, or because of illness or injury.

Determine the Dog's Pulse Rate and Character

Locate the femoral artery by placing the flat of your hand in the groin area and then gently press in on the middle of the inner thigh with the index and middle fingers until you feel pulsations.

Count the number of pulsations for 60 seconds or count for 30 seconds and multiply by 2 to determine pulses per minute. The normal pulse range is 70 to 120 pulses or beats per minute (bpm).

Judge the pulse character using the following terms:

- Regular or irregular rhythm.
- Strong or weak strength.
- The normal pulse character is regular and strong.

Determine the Dog's Respiratory Rate and Character

Count the number of times the dog breathes to determine breaths per minute by counting the number of breaths taken in 60 seconds or count the number of breaths taken in 30 seconds and multiply by 2. The normal respiratory rate of a dog is from 10 to 30 breaths per minute.

Judge respiratory character based on the depth (shallow, deep, or normal), the rhythm (panting, regular, or forced). The normal respiratory character of a dog is a normal depth and regular rhythm.

Determine the Dog's Mucous Membrane Color and Mucous Membrane Moistness

The best place to check mucous membrane color and moistness is the tissue covering the gums in the mouth.

Expose the dog's gums by gently pulling the top lip up or the bottom lip down and note the color of the gums. The normal mucous membrane color of a dog is pink. Pink mucous membranes tell us that enough oxygen is making it into the blood stream. Abnormal mucous membrane color would be pale, white, blue, yellow, or brick-red. Some breeds have black pigmented mucous membranes. If this is the case, place your thumb on the skin just under the lower eyelid and gently pull down and observe the color of the membranes of the inner lower eyelid.

Note the moistness of the gums by gently touching your finger to the exposed gums. Mucous membrane moistness is one of several crude assessments of hydration status of the dog. Normal mucous membranes are moist or slippery. Mucous membranes dry or tacky to the touch are not normal.

Determine the Dog's Capillary Refill Time (CRT)

CRT is the amount of time, measured in seconds, that it takes blood to return to an area of the gum after it has been blanched by your finger. CRT assesses blood flow to tissues.

Expose the dog's gums by gently pulling the top lip up or the bottom lip down. Gently press your index finger into the gums to blanch the area. Release the finger and count in seconds how long it takes for blood to return to the area. The normal CRT of a dog is less than 2 seconds.

Determine the Dog's Skin Elasticity

Skin elasticity is another of the crude assessment tools we have to evaluate hydration status of a dog.

Gently grasp a small area of skin on the back and pull it up into a "tent." Hold for a few seconds and then release. Note how long it takes the skin to return to normal. The normal skin elasticity in a dog is immediate return of tented skin to its normal position. Skin that remains tented more than 1 or 2 seconds is a crude indicator of dehydration.

Observe the Dog's Level of Consciousness, or Mental Alertness
Use one of the following terms to describe the dog's mental alertness:

- Bright, alert, and responsive (BAR), or quiet, alert, and responsive (QAR). The dog appears normal in all respects mentally.
- Depressed. The dog appears "down," lethargic, and not interested in normal activities (work, play), and may have a loss of appetite; the dog responds to verbal and physical stimuli, but is slow to respond.
- Stupor. The dog acts "drunk" and "out of it"; the dog responds to physical stimulation but not verbal stimulation; responses are very slow.
- Coma. The dog is completely unresponsive to verbal and physical stimulation.
- Agitated. The dog can't sit still, moves rapidly and irregularly, and acts "disturbed."

Determine the Dog's Weight and Body Condition Score (BCS)
Do this by weighing the dog on the scale and observing the dog's physical appearance. BCS should be determined utilizing the Purina™ Body Condition Score chart as a reference, located at purina.com/dogs/health/bodycondition.aspx. The optimal BCS for an MWD is a score of 4 or 5. Any MWD that is above or below the optimal BCS range is possibly over- or underweight.

Record Vital Signs Using the Following Format
- Body temperature: T–XXX.X° F.
- Pulse rate: P–XX bpm (beats per minute).
- Pulse character: Regular or irregular; strong or weak.
- Respiratory rate: R–XX breaths /min (or panting).
- Respiratory character: normal, shallow, or deep; regular, panting, or forced.
- Mucous membrane color: MM–color observed.
- Mucous membrane moistness: Moist or dry/tacky.
- Capillary refill time: CRT–≤2 (less than) or >2 (more than) seconds.
- Skin elasticity: normal or slow.

- Level of consciousness or mental alertness: BAR, QAR, depressed, stupor, coma, or agitated.
- Body weight: W–XX.X lbs. or XX.X #.
- Body Condition Score: X out of 9 or X/9.

Make note of any other significant observations or abnormalities. Be as specific as possible and notify the veterinary staff immediately of any abnormalities or significant findings.

PERFORM A PHYSICAL EXAM

To identify an illness or injury, you must recognize what is normal for your MWD. Sometimes the condition is so obvious that there is no question it is abnormal. Frequently, changes in your dog's health and disposition are subtle and it is important they are recognized. Early recognition of a serious problem can save your MWD's life.

Perform a Physical Exam of the Dog

- Note any external obvious signs of injury or illness.
- Prior to restraining your dog for physical exam, observe the animal in its natural state (that is, in the kennel or in the exercise yard). Look for things such as abnormal behavior, attitude, or level of consciousness; food and water intake; working or playing normally; vomiting or diarrhea; normal urination and defecation; or lameness or any other obvious signs of injury or illness.
- Measure and record the dog's vital signs (see page 1, "Take Vital Signs").
- Examine the dog's head looking for abnormalities including but not limited to eye discharge, nasal discharge, areas of hair loss, swellings, masses, sores, and obvious deformities.
- Evaluate the dog's eyes looking for foreign objects lodged in the eye, eye trauma or an eye out of its socket, masses, twitching or spasms, abnormal discharge such as blood or pus, and cloudiness of the clear part of the eye (cornea).
- Examine the dog's muzzle looking for obvious deformities, swelling, sores, and discharge.
- Examine the lips looking for obvious deformities, warts or similar bumps, masses, and redness or swelling.

- Remove the muzzle and examine the inside of the dog's mouth, if the dog will allow it. Look for obvious abnormalities such as broken teeth or a cut tongue, masses, cuts or sores, redness or swelling, foreign bodies, and abnormal odor.
- Examine the ears for foreign substances or debris, a dark, dry, waxy debris (signs of ear mites), bacterial or yeast infections produce a moist, greenish-yellow substance and an abnormal odor.
- Examine the dog's hair coat and skin, looking for areas of hair loss, parasites (lice, fleas, and ticks), redness and swelling, crusts, scales, masses, and matted areas.
- Examine the trunk and limbs by feeling the muscles and bones of the rib cage and front and hind legs. Note any swelling, masses, and pain response.
- Flex and extend all the joints of the front and hind legs and note any swelling and pain response.
- Check the spaces between toes of the paws looking for foreign objects, cuts and scrapes, wounds, swelling, or masses.
- Check the nails for proper trimming. Nails should not extend beyond the pad of the toes.
- Observe the genitalia. In both the male and female dog, look for inflammation, swelling, obvious deformities, abnormal discharge (a small amount of yellowish-green discharge from the prepuce is normal), and sores, especially of the scrotum in male dogs.
- Observe the rectum and anal area looking for inflammation, swelling, masses, sores, or wounds.
- Examine the tail looking for wounds, sores, and areas of hair loss.
- Make a note of all observations recording vital signs as trained in "Motivation" (previous section), and specific information concerning any abnormalities.
- Notify the veterinary staff immediately of any abnormalities or significant findings.

PERFORM A PRIMARY SURVEY

Military Working Dogs can become seriously ill or injured in a very short period of time. It is critical to identify life-threatening problems immediately. A Primary Survey is a rapid examination that is designed to target the most critical body systems in order of importance to detect serious problems. The survey should be done quickly (less than two minutes).

Visually Assess the Dog from a Distance as You Approach It

Note the level of consciousness, responsiveness, and any unusual behavior or activity. Note unusual body or limb postures or positions that suggest bone fractures, joint dislocations, or other traumatic injuries. Listen for unusual breathing sounds and for any audible airway obstructions. Look for obvious blood, wounds, or other gross abnormalities. Immediately notify the kennel master or veterinary personnel if any abnormalities are noted.

Assess the Airway by Listening for Labored and Noisy Breathing that Suggests Something Is Blocking the Airway

Feel the throat area and trachea (wind pipe) in the front part of the neck. Look for obvious masses, wounds, swellings, or deformities that may cause airway obstruction. If possible, open the mouth and examine the inside and as far back into the throat area as you can see. Look for masses, foreign objects, swelling, or deformities that may cause airway obstruction. If possible, clear airway obstructions using a "finger sweep" technique (run index and middle fingers along the cheek and across the back of the dog's throat) to remove any large objects or blood clots or vomit.

Assess Breathing by Watching the Dog for Clues to the Location of Lung or Airway Trauma or Problems

Deep, labored breathing suggests lung trauma or lung problems, such as lung bruising. Shallow, rapid breathing suggests air, blood, or some other fluid in space around the lungs inside the chest cavity. If the dog is not breathing, it is in respiratory arrest; this is an emergency condition and you should seek immediate assistance from the vet. Irregular breathing may indicate brain injury. Look at the mucous membranes (gums). Blue, pale or white, yellow, or bright red gums are abnormal. Feel the chest rise and fall with each breath. If breathing is not effective, immediately contact veterinary personnel and request further guidance. Be prepared to provide basic cardiopulmonary life support if the dog is not breathing or not breathing well.

Assess Circulation by Determining the Dog's Pulse Rate and Character (in accordance with "Take Vital Signs" in "Motivation" section above)

A very slow pulse rate or a very rapid pulse rate suggests major trauma or medical problem. Absence of pulse rate indicates cardiac arrest. Be prepared to provide basic cardiopulmonary life support if the dog's pulse rate is less than 60 beats per minute, there is no pulse detected, or if the pulse is weak or irregular. Determine the dog's CRT. Prolonged CRT (> 2 seconds) suggests poor blood flow to tissues. Be prepared to provide intravenous fluid therapy for shock or basic cardiopulmonary life support if the CRT is prolonged.

If you determine there is a problem with the heart or circulation, immediately contact the kennel master or veterinary personnel and request further guidance.

Perform a Brief, Rapid Examination of the Rest of the Dog

Quickly assess the dog's body for wounds, fractures, and evidence of trauma elsewhere (painful areas, swelling, bruising, skin abrasions). Pay particular attention to the spinal column, abdominal region, flank, and limbs for signs of trauma.

If necessary, evacuate the MWD to the nearest veterinary facility. Make a written record of the treatment, including the date, time, and actions taken.

PROVIDE FIRST AID FOR A BLEEDING WOUND

Uncontrolled bleeding can be fatal or cause shock and lead to further complications. Serious bleeding, especially arterial bleeding, must be controlled immediately.

Venous Bleeding (Bleeding from Injured Veins)

Venous bleeding is generally less likely to cause shock or death unless major veins are injured. Venous bleeding is more likely in skin wounds, lower leg wounds, paw wounds, and face and neck wounds. Venous bleeding is usually dark in color and usually oozes from the injury site. First aid for most venous bleeding involves applying immediate direct pressure and a pressure bandage.

Arterial Bleeding (Bleeding from Injured Arteries)

Arterial bleeding is much more likely to cause shock and death, and must be managed more aggressively than venous bleeding. Arterial bleeding is more likely in groin and armpit wounds, deep neck wounds, and leg and paw wounds. Arterial bleeding is usually bright red in color and usually spurts or flows rapidly from the injury site. First aid for arterial bleeding requires immediate direct pressure followed by application of a hemostatic clotting agent and application of a pressure bandage.

Providing First Aid for Mild Bleeding

- Immediately apply pressure with your hand and continue to hold firm pressure while you or another person gathers your first aid supplies.

- Apply 5–10 sterile 4x4 gauze sponges to the bleeding wound. If sterile 4x4 gauze sponges are not available, use clean pieces of cloth, a field dressing, or similar material. The key is to control bleeding; dirty wounds and infections can be dealt with later.

- Continue to apply firm pressure to the wound with the bandage between the wound and your fingers.

- Using direct pressure to stop bleeding takes time. DO NOT lift the bandage or remove the bandage to look at the wound because this will break up the clot that is forming and bleeding will begin again.

- If the bleeding leaks through the gauze or cloth you applied, APPLY more gauze or cloth; DO NOT remove the original gauze or cloth.

- Without removing the gauze sponges, apply a bandage to provide direct pressure and control bleeding. This allows you to do other things, such as coordinating a medical evacuation (MEDEVAC).

- Wrap the bleeding wound with 1–4 rolls of roll gauze. Usually, lower leg wounds require 1–2 rolls, and higher limb wounds and body wounds require 4 rolls. The roll gauze should be applied tightly to provide pressure to the bleeding wound. Wrap the area with 1–3 rolls of elastic conforming bandage.

- If your first aid kit is not available, use whatever it is you have to apply a protective bandage with pressure over the bleeding site. Field dressings, a cut or torn T-shirt, or cloth material can be used. Either use medical adhesive tape to secure the bandage, strips of cloth, or the field dressing tapes to tie the bandage in place.

Providing First Aid for Moderate or Severe Bleeding

- Immediately apply pressure with your hand and continue to hold firm pressure while you or another person gathers your first aid supplies.
- Apply 1 full packet of the hemostatic clotting agent from the first aid kit directly into the wound.
- Immediately cover the wound with 10–15 sterile 4x4 gauze sponges as for mild bleeding. Continue to apply firm pressure to the wound with the bandage between the wound and your fingers.
- Using direct pressure to stop bleeding takes time. DO NOT lift the bandage or remove the bandage to look at the wound because this will break up the clot that is forming and bleeding will begin again.
- If the bleeding leaks through the gauze or cloth you applied, APPLY more gauze or cloth; DO NOT remove the original gauze or cloth.
- Without removing the gauze sponges, apply a bandage as described for mild bleeding to provide direct pressure and control bleeding.
- Observe for signs of pain and discomfort. If the bandage is too tight, it may interfere with circulation to the point of requiring an amputation.

Inform the kennel master of the situation and immediately contact the closest veterinary staff and request further instructions.

Make a written record of the treatment, including the date, time, and actions taken.

PROVIDE FIRST AID FOR UPPER AIRWAY OBSTRUCTION

Recognizing signs of an upper airway obstruction is imperative. Typically a dog playing with or chewing on an object, followed immediately by pawing at its face or throat, acting frantic, trying to cough and choke, with sudden onset of difficulty breathing with abnormal "snoring" breathing sounds is a good indication the MWD's airway is blocked.

Upper Airway Obstruction Is a Life-Threatening Situation

You must perform first aid as quickly as possible. **The following steps should be completed in less than 30 seconds.** The dog may or may not have lost consciousness.

1. Determine that the dog has an upper airway obstruction by checking the airway.
2. Gently tilt the head slightly back and extend the neck.
3. Look in the mouth and identify anything that is blocking the airway, such as vomit, a ball, a stick, clotted blood, bone fragments, or other object.
4. Use a gauze sponge to grasp the dog's tongue and pull it forward to improve visualizing the mouth.
5. If you are able to visualize a foreign object use the "2 finger sweep" technique to remove fixed objects. Run your index and middle fingers into the dog's mouth along the cheek and across the back of the throat, removing any foreign objects that are visualized or felt. If unable to dislodge the foreign object, immediately transport to the nearest veterinary treatment facility as the airway may have become blocked due to swelling.
6. You may also use the modified Heimlich maneuver to remove mobile foreign objects such as a tennis ball or Kong.
   • Grasp the dog around the waist so that the rear is nearest to you, similar to a bear hug.
   • Place a fist just behind the ribs.
   • Compress the abdomen several times (5 times) with quick thrusts.
   • Check the mouth to see if the foreign object is dislodged.
   • Repeat the modified Heimlich maneuver 1–2 times if initial efforts are unsuccessful.
   • If unable to dislodge the foreign object, immediately transport to the nearest veterinary treatment facility as the airway may have become blocked due to swelling.
7. If the dog has lost consciousness and you were able to successfully remove the foreign object, the dog may regain consciousness on its own or cardiopulmonary resuscitation (CPR) may need to be performed. Immediately perform a primary survey and take appropriate action.

Report the event by immediately contacting supporting veterinary personnel for further instructions and notify the kennel master.

Even if you are successful in removing a foreign object, veterinary examination is required. Internal injury could have occurred that may not be evident.

Make a written record of the treatment, including the date, time, and actions taken.

PERFORM CARDIAC ARREST LIFE SUPPORT

Assess whether the dog is in cardiopulmonary arrest, cardiac arrest, or respiratory arrest **within 30 seconds.** Use the mnemonic **"CAB"** to focus your attention on the **c**irculation, **a**irway, and **b**reathing as you assess the MWD.

- Cardiac arrest is determined when the heart has stopped beating, no pulse can be found but the dog is breathing voluntarily.
- Respiratory arrest is determined when the dog is not breathing voluntarily, but the heart is beating and a pulse is present.
- Cardiopulmonary arrest is determined when the heart has stopped beating, no pulse can be found, and the dog is not breathing.
- Try verbally and physically to get the dog to respond. If the dog responds, it does not need basic cardiac life support (BCLS).
- If the dog is unresponsive, immediately call for help if others are nearby. Have someone request support from veterinary personnel. Although BCLS requires at least two people to be most successful, continue with the following steps even if you are alone.

Check for Blood Circulation
- **Look** at the dog's gums to assess the color of the mucous membranes and CRT.
- **Listen** to the chest for a heartbeat.
- **Feel** for a pulse at the femoral artery.

Clear the Airway
1. Gently tilt the head lightly back and extend the neck.
2. Look in the mouth and remove anything that is blocking the airway, such as vomit, a ball, a stick, clotted blood, bone fragments, or other objects.

Check for Breathing
1. **Look** for the rise and fall of the chest.
2. **Listen** to the dog's mouth and nose for signs of breathing.
3. **Feel** breath on your skin by placing your face or hand near the dog's mouth and nose.

Take Action Based on Your Findings
- If the dog is not breathing, but has a pulse or heart rate, the dog is in respiratory arrest. Begin rescue breathing immediately.
- If the dog has no pulse or heart rate, the dog is in cardiac arrest. If the dog is not breathing voluntarily and has no heart beat or pulse, the dog is in cardiopulmonary arrest. Begin BCLS immediately.
- Be very careful not to get bitten! Even if the dog would not normally bite you, the dog may not have normal control of its actions. If your dog is conscious, it is NOT in cardiopulmonary arrest and does NOT need BCLS.
- Perform assisted BCLS **within 2 minutes** of determining the dog has no pulse or heart beat.

Determine with your assistant who will give chest compressions and who will give "mouth-to-snout" breathing. BCLS on a large dog is physically demanding work. Be prepared (by practicing) to rotate positions with other personnel with minimal interruption of chest compressions and rescue breathing.

Position the dog
1. Kneel next to the dog.
2. Place the dog on its side (lateral recumbency) with its spine against your body.
3. Bend the dog's front leg up so the elbow moves about ⅓ of the way up the chest; release the elbow and make a note of the area, as this is the spot to place your hands to perform chest compressions.

Position your hands
Position your hands by placing one hand on top of the other with all fingers closed together. Place your hands on the chest wall at the position you identified above.

Perform chest compressions
1. With partially locked elbows, bend at the waist and apply a firm, downward thrusting motion.
2. Compress the chest wall approximately 6 inches at a sustained rate of 100 compressions per minute, which is about 1 compression every half-second. **Proper chest compressions are the most important part of BCLS.** Do not stop chest compressions to direct or assist in other actions unless safety is an issue.

Clear the airway to remove upper airway obstructions and open the airway for better rescue breathing by removing the dog's collar, pull the tongue out in downward motion using gauze to hold onto it. Visually inspect the inside of the dog's mouth for obstructions and feel along the outside of dog's throat for obstructions. Using either the 2 finger sweep technique or the modified Heimlich maneuver, remove obstructions if possible.

Perform rescue breathing
Use the "mouth-to-snout" method within 30 seconds of clearing the airway.

1. Seal the dog's mouth and lips by placing your hands around the lips, gently holding the muzzle closed.
2. Place your mouth over the dog's nose and forcefully exhale into the nose.
3. Give 2 quick breaths first, then check to see if the dog is breathing without assistance.
4. If the dog does not breathe voluntarily, continue breathing for the dog at a rate of 20 breaths per minute (one breath every 3 seconds).

Check the dog's response after 4 minutes of BCLS, and every 4 minutes thereafter.

Check for voluntary breathing and for a heartbeat or pulse. If there is no voluntary breathing or a heartbeat or pulse, continue BCLS. If there is a heartbeat or pulse, but no voluntary breathing, stop chest compressions but continue rescue breathing. Every time BCLS is stopped, blood pressure drops and blood flow and ventilation stop. Frequent stopping results in poor survival rates. Stop only every 4 minutes and only long enough to quickly check the patient's breathing, pulse, and heart beat.

Perform UNASSISTED BCLS within 2 minutes of determining BCLS is necessary
1. Kneel next to the dog, position the dog, and perform chest compressions exactly as you do for assisted BCLS.
2. Perform "mouth-to-snout" breathing 2 times after every 15 chest compressions.
3. Maintain chest compressions and rescue breathing at a compression –breathing cycle of 15 compressions–2 breaths.
4. Check the dog's response every 4 minutes (gently tilt the head slightly back and extend the neck; look in the mouth and remove anything that is blocking the airway).
5. Continue BCLS as long as the dog does not have a pulse or heart rate or is not breathing on its own.

Discontinue BCLS under the following circumstances
1. The dog is successfully resuscitated (has a pulse and heartbeat and is breathing on its own).
2. The dog has not been resuscitated after at least 20 minutes of BCLS.
3. You are directed to stop BCLS by a more senior handler, kennel master, or veterinary personnel.

Report your actions and initiate MEDEVAC
Immediately contact supporting veterinary personnel for further instructions, notify the kennel master, and make a written record of the treatment, including the date, time, and actions taken.

PROVIDE FIRST AID TO MWD WITH AN ALLERGIC REACTION

Check with local resources to identify venomous snakes, arthropods, and insects in your area that could cause potential harm to your dog. Reliable resources would be the kennel master, local community health center's preventive medicine department, and supporting veterinary personnel.

It is best to know the venomous snakes and insects by sight or characteristic markings. After getting information about the snakes and insects, try to commit to memory their habits and behavior.

Recognize the signs of an allergic reaction to envenomation by an insect, arthropod, or snake. Mild signs may include any of the following:

- Apparent pain at the wound site: The dog may lick or bite the area; if the wound is to a paw or leg, it may hold it up and not put any weight on it.
- Fang marks, bite marks, or puncture wounds.
- Drops of blood or oozing blood at the wound site.
- Swelling.
- Excessive salivation (drooling).

Severe signs may include any of the following:

- Weakness, lethargy, disorientation.
- Muscle tremors.
- Slow, labored breathing.
- Vomiting.
- Diarrhea.
- Tissue necrosis (death) with open, draining wounds at bite site.
- Collapse or unconsciousness.
- Shock, including pale or blue mucous membranes, weak or absent arterial pulse, prolonged CRT, collapse, increased heart rate.
- Death.

Not all snake and insect bites or stings will cause an allergic reaction. Your dog may not suffer an allergic reaction to a bite or sting, but still may require immediate treatment for envenomation by veterinary staff. The severity of symptoms your dog displays are based on the amount and type of venom injected through the bite or sting, the location of the bite or sting, and the size of the dog.

Typically, an allergic reaction is immediate, occurring within 5 minutes of envenomation. If there are no visible signs of a reaction after an hour, then it is unlikely that an allergic response to envenomation occurred.

You may or may not witness your dog being bitten or stung. However, if you see your dog bitten by an insect, arthropod or snake, take immediate action and provide first aid.

Provide First Aid for an Allergic Reaction to Insect, Arthropod, or Snake Envenomation

If the signs have progressed so much that the dog has ceased breathing, its heart has stopped beating, or shock is present, take immediate action to treat for these problems.

Administer drugs to reduce the allergic reaction. Give one (1) intramuscular dose of diphenhydramine (50 mg/ml), using the following dose chart.

Table 1. Diphenhydramine Dosage Chart

Body weight (in pounds)	Volume of diphenhydramine to give intramuscularly (in milliliters)
30 to 35	0.7
36 to 40	0.7
41 to 45	0.9
46 to 50	1.0
51 to 55	1.1
56 to 60	1.2
61 to 65	1.3
66 to 70	1.4
71 to 75	1.5
76 to 80	1.6
81 to 85	1.7
86 to 90	1.8
91 to 95	1.9
96 to 100	2.0
101 to 105	2.1
106 to 110	2.2
111 to 115	2.3
116 to 120	2.4

Give one (1) intramuscular dose of dexamethasone sodium phosphate (4 mg/ml), using the following dose chart.

Table 2. Dexamethasone Sodium Phosphate Dosage Chart

Body weight (in pounds)	Volume of dexamethasone sodium phosphate to give intramuscularly (in milliliters)
30 to 35	1.8
36 to 40	1.9
41 to 45	2.4
46 to 50	2.7
51 to 55	3.0
56 to 60	3.3
61 to 65	3.6
66 to 70	3.9
71 to 75	4.1
76 to 80	4.4
81 to 85	4.7
86 to 90	5.0
91 to 95	5.3
96 to 100	5.6
101 to 105	5.9
106 to 110	6.1
111 to 115	6.4
116 to 120	6.7

Keep the dog calm and quiet; cease operations with the dog. If possible keep the affected area lower than the heart. Inform the kennel master of the situation, and contact supporting veterinary staff to request further instructions.

If the dog's condition deteriorates, initiate a MEDEVAC. If an open wound is present or develops, protect the wound with a bandage.

DO NOT do any of the following, as these make the allergic reaction worse:

- Apply ice to the bite or sting area.
- Exercise or have the dog move around. Movement causes the venom to spread more quickly.
- Apply a tourniquet if the bite or sting was to an extremity.
- Cut the area to squeeze or suction out the poison. Note: Rattlesnake anti-venin is the only definitive treatment for snakebite; this is only available from your supporting veterinary personnel. Specific treatments are going to vary with the type of snake, spider, or insect involved.
- For true anaphylactic reactions (collapse, increased heart rate, weak or absent pulses), initiate treatment for shock.

Make a written record of the treatment, including the date, time, and actions taken.

PROVIDE FIRST AID FOR DEHYDRATION

Dehydration is the excessive loss of fluids and electrolytes from the body through illness or physical exertion. Electrolytes (sodium, chloride, potassium) are salts needed by cells to control movement of water in the body and to control many body functions. Understand the definition of dehydration and common causes in MWDs.

Causes of dehydration include inadequate water intake or loss of water and electrolytes due to illness (fever, diarrhea, and vomiting) and environment (heat, humidity, and cold).

Determine that the Dog Is Dehydrated by Observing Signs of Dehydration

The early signs of dehydration are very hard to recognize. You must know your MWD well to identify them.

Signs of early dehydration

- Reduced physical activity.
- Abnormal mental activity or level of consciousness (depressed, lethargic).
- Tacky gums (mucous membranes) and a dry nose.

Signs of moderate dehydration

- Dry and tacky mucous membranes (nose, mouth, gums).
- Loss of skin elasticity/increased skin elasticity—the skin doesn't snap right back to place as it normally does.
- Slightly sunken eyes.
- Slightly increased CRT.

Signs of severe dehydration

- Pale mucous membranes.
- Prolonged CRT.
- Weight loss (5% or more).
- Sunken eyes.
- Weak arterial pulse.

Provide First Aid for Dehydration

If your dog is showing early signs of dehydration, offer fresh water. Unfortunately, if your dog is already dehydrated, sick, injured, or cold, it may not want to drink.

If the dog does show an interest in drinking water, make sure the dog doesn't drink more than a few sips every few minutes. Overdrinking, or drinking quickly, could lead to vomiting, dehydrating the dog further.

If the dog is vomiting, has diarrhea, is showing signs of heat injury or signs of moderate to severe dehydration, contact veterinary personnel immediately for possible emergency treatment.

If the environment is hot, humid, or sunny, move the dog to shade or indoors if air conditioning is available and allow the dog to rest.

If the dog is showing signs of moderate dehydration, administer 1 liter of Lactated Ringer's solution (LRS) fluids (beneath the skin) in 4 separate locations. If the dog is showing signs of severe dehydration, administer 1 liter of LRS fluids over a 2-hour period intravenously.

For all dogs with dehydration, monitor for shock and provide appropriate first aid if signs of shock develop. Evacuate the MWD to the nearest veterinary facility if it is showing signs of severe dehydration or shock. Make a written record of the treatment, including the date, time, and actions taken.

ADMINISTER SUBCUTANEOUS FLUIDS

Administering subcutaneous fluids is a method to provide water and electrolytes for dehydrated MWDs. Understand reasons for use of subcutaneous fluids.

Subcutaneous fluid administration is acceptable only for MWDs with mild dehydration due to inadequate water intake or excessive loss of body water and electrolytes (illness or environmental factors).

If moderate or severe dehydration is present, or if shock is present, use other methods for fluid administration. Subcutaneous fluids are not appropriate for dogs exhibiting signs of severe dehydration or shock.

Recognize Clinical Signs of Mild Dehydration or Historical Facts that Suggest Dehydration Is Present

- Abnormal mental activity (depressed, lethargic).
- Decreased performance.
- Prolonged skin elasticity (prolonged skin "tenting").
- Prolonged CRT.
- Tacky mucous membranes (gums).
- Slightly sunken eyes.

Historical Facts that Suggest Dehydration Is Present

- Three or more episodes of vomiting or watery diarrhea in the past 24 hours.
- Moderate or heavy work in hot and/or humid environment.
- Recent illness with decreased water intake.

Assemble Supplies and Prepare Equipment for Use

- One 1-liter bag of sterile LRS.
- Fluid administration set.
- Four 18-gauge needles.
- four to six 4x4 gauze sponges.
- Isopropyl alcohol.

Prepare Equipment

1. Remove the wrapper from the bag of LRS.
2. Remove the administration set from its packaging.
3. Close the flow regulator on the administration set. You must roll the flow regulator toward the end of the line to close it. This pinches the tubing closed so it won't leak.
4. Remove the cover from the injection port of the bag of LRS.
5. Remove the cap from the administration set spike and insert the spike into the injection port of the fluid bag. Maintain sterility at all times and avoid contaminating uncapped surfaces. Don't allow any uncapped surfaces to come in contact with anything.
6. Hang the bag or have someone hold the bag two to three feet above the dog. Put the bag under your armpit if there are no other options.
7. Squeeze the sides of the administration set chamber several times to force fluid into the chamber. Fill the chamber half way or to the arrow mark or line on the side.
8. Remove one 18-gauge needle from its outer packaging. Do not touch the exposed end of the needle.
9. Remove the cap from the drip set line and attach the needle to the end. Twist the needle and make sure it's seated snugly on the line. Avoid contaminating either the needle or the line.
10. Move the flow regulator back to the open position to allow the fluid to flow freely out the end of the line. Let fluid flow out until all air bubbles are gone from the line. Once the air bubbles are gone, stop the flow of fluid by rolling the regulator back down. Recap the exposed end.
11. Visually divide the bag of fluids into four equal parts, of about 250 ml each. You will be administering approximately one quarter of the bag (250 ml) in each subcutaneous location. Use the markings on the fluid bag to determine how much fluid you will administer to each site.

Administer Subcutaneous Fluids

Select an area of skin. You will be injecting fluids into the skin in four separate places, so you will need to select four areas.

Choose parts of the body where there is loose skin such as over the shoulder blades (above the front legs) and over the rib cage on each side.

1. Soak a 4x4 gauze sponge with isopropyl alcohol and vigorously scrub the skin over the injection site to remove dirt and skin oils.
2. Pinch up a fold of skin forming a "tent" or inverted V.
3. Uncap the needle from the fluid administration set, and insert the needle quickly and firmly into the center and toward the bottom of the tent at a horizontal angle. The needle should go in easily and should not hit any obstructions.
4. The needle should go in about ½" to ¾" and should not go through the other side of the tent.
5. Release the skin while still holding onto the needle.
6. Roll the flow regulator to the open position, allowing fluid to flow through the tube and into the dog. It is normal to see large bumps appear in the areas when you are administering fluids. These are fluid pockets. These bumps will last for several hours as the fluid is slowly absorbed.
7. Administer one quarter (250 ml) of the bag of fluids at the first site.
8. Mark the LRS bag.
   - Tear a piece of tape approximately 12" long.
   - Place the tape on the bag vertically along one side next to the volume marks printed on the bag by the manufacturer.
   - Make marks on the tape to help you control how fast you give the fluid.
   - At the prescribed rate of administration (1 liter per hour for 2 hours), you will give 250 ml of fluid every 15 minutes.
   - Make marks on the tape at the 250 ml, 500 ml, 750 ml, and 1000 ml lines on the bag. These marks correspond to 15 minutes, 30 minutes, 45 minutes, and 60 minutes of time.
   - You will use the flow control dial to give the correct amount of fluid over time.
   - Mark a start line on the tape at the point where the fluid level and the tape meet. This is the start line. It should be at about the 1000 ml mark.
9. Stop the flow with the flow regulator when the appropriate amount of fluid is given.

10. Carefully remove the needle from the skin and apply pressure to site for a few seconds to prevent bleeding or fluid leakage.
11. Carefully recap the needle.
12. Change the needle and replace the used needle with a sterile needle.

Repeat steps 1 through 11 three additional times at three separate sites, administering approximately one quarter (250 ml) of the bag at each site.

Contact supporting veterinary personnel for further instructions. Make a written record of the treatment, including the date, time, and actions taken.

PROVIDE FIRST AID FOR SHOCK

Understand Shock and Common Causes of Shock in MWDs
Shock is the body's response to a traumatic injury or severe illness in which blood flow to vital organs like the brain, heart, lungs, liver, and kidneys is not adequate for survival because of lack of oxygen delivery to these organs.

- Shock is progressive, meaning that if not treated quickly, shock may worsen. Even with effective treatment, shock can ultimately cause death.
- Shock is a life-threatening situation. Emergency first aid must be provided immediately to improve chances of survival.

In most cases of shock, fluid therapy is one of the most important treatment measures aimed at improving blood flow.

Common Causes of Shock in MWDs
- Trauma with blood loss (motor vehicle accident, gunshot injury, shrapnel wound, blast injury).
- Severe dehydration (vomiting, diarrhea).
- Heat injury.
- Allergic reactions to snake bites and insect stings.
- Poisoning.

Shock Is Always Caused by Something Else

It is important to identify if possible the primary cause, because first aid for the primary cause is just as important as first aid for shock. You may have to provide first aid for shock at the same time you are providing first aid for the primary cause.

1. Perform a primary survey of the MWD to identify a primary cause for the shock.
2. Treat any other severe injuries such as respiratory arrest, cardiopulmonary arrest, arterial bleeding, heat stroke, allergic reaction to snake bite or insect sting, poisoning, severe dehydration or Gastric Dilatation–Volvulus Syndrome (GDV or bloat) while preparing to treat the shock.

Recognize Signs of Shock

- Panting or labored breathing.
- Bright red mucous membranes (gums), or pale, gray, or blue mucous membranes.
- Increased heart rate.
- Weak or absent arterial pulse.
- Prolonged CRT >2 seconds.
- Low body temperature.
- Cold paws.
- Weakness, collapse.
- Depressed, acting "out of it," lethargic, coma.

Provide First Aid to Treat Shock

1. Place an 18-gauge intravenous catheter in a leg vein and administer intravenous fluids.
2. Give 1 liter of LRS each hour for 2 hours. Do not give more than 1 liter in an hour or more than 2 liters total.
3. If oxygen is available, use a face mask or tubing and blow oxygen into the dog's nose or mouth at a flow rate of 5-10 liters per minute.

4. If you must move the dog, do so very gently and try to keep the dog on a flat surface. The dog may have a spinal cord injury that is not obvious.
5. Cover the dog with something to keep it warm (e.g., blanket, towel, clothing, etc.).
6. Try to keep the dog calm by speaking calmly and reassuringly to it.

Monitor the Dog and Your Treatment
1. Continue to monitor vital signs and note whether the dog is responding to treatment.
2. Monitor the intravenous fluid (IV) catheter insertion site on the leg. If swelling is noted around the catheter or if the catheter is not working or becomes blocked, place a second intravenous catheter in another leg vein and use the new catheter.
3. Monitor the intravenous fluid therapy equipment. Look for:
   • Disconnects anywhere in the setup, kinks or blocks in the tubing, and monitor fluid administration rate.
   • Use the marks you created on the tape/bag and a timepiece to ensure the administration rate is correct. Adjust the drip rate by rolling the regulator up or down if the fluids are flowing too fast or too slow.

Continue to provide first aid for any medical problems found on the primary survey.

Make a written record of the treatment, including the date, time, and actions taken.

PROVIDE FIRST AID FOR HEAT INJURY

Heat Injuries Result When the Body's Natural Cooling Mechanisms Fail in Response to Internal Overheating
Humans regulate body temperature mostly by sweating. Because dogs do not sweat, they regulate their body temperature by panting. If a dog is unable to cool enough, its internal temperature will rise and the dog may progress through different stages of heat injury.

Normal body temperature for a dog is 100.5°F to 102.5°F. As the dog's temperature rises to 105° or 106°F, the dog develops heat stress.

If cooling measures are not taken immediately, the dog's body temperature will continue to rise, progressing to heat exhaustion, usually with a temperature of 106° to 108°F. Once the dog's body temperature rises over 108° F, heat stroke is likely. All phases of heat injury are life-threatening situations.

The progression of heat injuries can be quite rapid, sometimes only taking a few minutes. Occasionally, there is little or no warning and the progression is so rapid the dog might already be suffering heat stroke when the situation is discovered.

Short-term effects on a dog suffering from a heat injury may include shock, organ damage, or death. Long-term effects may include brain damage and the increased possibility of recurrence of a heat injury.

Immediate first aid is required for any MWD with suspected heat injury.

Causes of Heat Injury Are Divided into Environmental Causes and Exertional Causes

Environmental causes are due to exposure to high environmental temperature and humidity or a combination of both; confinement to a small, hot space such as a crate, kennel run, or vehicle; poor acclimation to heat and humidity; or inadequate water intake.

Exertional causes are due to increased body temperature that develops with strenuous exercise or work (worsened by hot, humid, or hot and humid environments), existing or undiagnosed disease or illness, medications or drugs, age, and previous heat injury.

The combination of exposure to high environmental temperatures and exertion can markedly increase the risk of heat injury.

Heat Injury Prevention

MWDs rely on their handlers to take care of them and make decisions for them, especially in extreme environments. Take care of, monitor, and know your dog!

Acclimation

Follow the human rules of acclimation. Increase workload and exposure to the environment gradually over a 14-day period. Use an acclimation period if the dog is recovering from an illness, as well.

Hydration
1. Make sure your dog is properly hydrated by allowing for frequent water breaks.
2. Ensure your dog always has access to water whether the dog is in its kennel, exercising, or resting.
3. Initially, dehydration is undetectable. By the time you are able to detect your dog is dehydrated, it is already too late.
4. Dehydration makes your dog more susceptible to heat injury and causes decreased performance.
5. Allow your dog to drink when you drink—typically a small amount every 10–15 minutes in an extreme or high activity environment.
6. Use fresh water. Do not give your dog "Cool Blue," Gatorade®, Pedialyte®, or similar fluids.

Fitness
Dogs need a physical training program to remain in shape. Out-of-shape dogs are more prone to heat injuries.

1. DO NOT confine an MWD in a small, poorly ventilated, hot area at any time.
2. Use the human heat category work/rest cycles and Wet Bulb Globe Temperature (WBGT) guide. The local supporting preventive medicine team at your installation will keep this information updated and available. Additionally, if the dog has had a prior heat injury, consider ceasing exercise when the temperature reaches 90°F. For all other dogs, consider ceasing exercise when the temperature reaches 95°F.
3. Do not use muzzles unless required for safety reasons. Loosen muzzles when possible to allow the dog to pant easier. A dog's cooling mechanism is its ability to pant.

Recognize Signs of Heat Injury
Mild heat injury ("heat stress") can be recognized by heavy, controlled panting; high rectal temperature, usually 105° to 106°F; fast, strong pulse; and slightly decreased performance. **NOTE:** Controlled panting means the dog can stop panting when an alcohol-soaked gauze is put in front of its nose or the dog becomes interested in something.

Uncontrolled panting means the dog cannot stop panting even when offered a treat or exposed to alcohol-soaked gauze.

Moderate heat injury ("heat exhaustion") can be recognized by very high rectal temperature, usually 106° to 108°F; uncontrolled panting; fast, strong or weak pulse; failure to salivate; tacky or dry nose and mouth; unwillingness to work or exercise; acts tired; loss of appetite; becoming unresponsive to handler and commands; staggering; weakness; depressed; acting "out of it"; or bright red mucous membranes (gums).

Severe heat injury ("heat stroke") can be recognized by extremely high rectal temperature, usually over 108°F; body is hot to the touch; vomiting; pale mucous membranes (gums); abnormal mental activity or level of consciousness—completely "out of it"; seizures; coma; diarrhea, sometimes with blood (bright red blood or dark, tarry feces); shock; or death.

Provide First Aid for Heat Injury

First aid for mild heat injury ("heat stress") is to cease working the dog and to cool the dog externally. Immediately cool the dog using one or more of the following methods.

1. Spray or pour cool water on the dog, or use soaked wet towels. DO NOT use ice or ice water, because this causes a serious rapid decrease in body temperature and results in hypothermia (dangerously LOW body temperature).
2. Move the dog to a shaded area if outdoors or into a cool building.
3. Circulate cool air near the dog using fans.
4. Loosen the muzzle and collar. Remove these if possible and safe.
   - Monitor and treat for shock, if shock develops.
   - Monitor vital signs every 5 minutes.
   - Discontinue cooling efforts when the rectal temperature reaches 103ºF.

First aid for moderate and severe heat injury ("heat exhaustion" and "heat stroke") is to cease working the dog.

- Cool the dog as for mild heat injury.
- Initiate intravenous fluid therapy.

1. Give 1 liter of LRS intravenously every hour for 2 hours. Do not give more than 1 liter of fluids every hour or more than 2 liters of fluid total.
2. Monitor the MWD and fluid therapy as directed.
3. Monitor and treat for shock, if shock develops.
4. Monitor vital signs every 5 minutes.
5. Discontinue cooling efforts when the rectal temperature reaches 103°F.

Notify the kennel master of the situation and contact supporting veterinary personnel.

ADMINISTER INTRAVENOUS FLUIDS

Prepare and Assemble Supplies for an Intravenous Infusion

Assemble supplies
- One roll of 1" medical adhesive tape.
- Two 1-liter bags of LRS.
- Fluid administration set.
- 18-gauge x 1½ inch intravenous catheter.
- Injection port adapter.
- One roll of self-adhesive conforming tape.

Prepare equipment
1. Maintain sterility at all times, especially avoiding contamination of uncapped surfaces.
2. Open the package of the intravenous catheter. Inspect the catheter to ensure serviceability (no burs, sterile, etc.). Flush with sterile LRS. Replace the catheter cap.
3. Flush catheter injection port with sterile LRS.
4. Tear three 12" strips of medical adhesive tape. Fold about ¼ of the end of each on itself to create a tab.
5. Open a package of self-adhesive conforming tape.
6. Remove the wrapper from the bag of LRS.
7. Remove the fluid administration drip set from its packaging.

8. Close the flow regulator on the tubing attached to the drip set. You must roll the flow regulator toward the end of the line to close it. This pinches the tubing closed so it won't leak.
9. Remove the cover from the injection port of the bag of LRS.
10. Remove the cap from the administration set spike and insert the spike into the injection port of the fluid bag.
11. Hang the bag or have someone hold the bag 2–3 feet above the dog. Hold the bag under your armpit if there are no other options.
12. Fill the drip chamber halfway by squeezing the sides of the chamber several times. If you overfill the chamber, just flip the bag upside down and squeeze the drip chamber several times to force fluid back into the bag.
13. Remove the protective cover from the end of the administration set line. Set the cap aside and keep it uncontaminated.
14. Remove air bubbles from the line by moving the flow regulator back to the slightly open position. This will allow fluid to flow slowly out of the end of the line. Let fluid flow out until all air bubbles are gone. Once the air bubbles are gone, stop the flow of fluid by rolling the regulator back down. Recap the exposed end. It is important to remove all air from the line so that air does not enter the dog's bloodstream. Air in the bloodstream can potentially travel to the heart and cause cardiac arrest. Do not allow too much of the fluid out while clearing the line. Allow just enough out to remove air bubbles.

Mark the LRS bag.

1. Tear a piece of tape approximately 12" long.
2. Place the tape on the bag vertically along one side next to the volume marks printed on the bag by the manufacturer.
3. Make marks on the tape to help you control how fast you give the fluid.
4. At the prescribed rate of administration (1 liter per hour for 2 hours), you will give 250 ml of fluid every 15 minutes.
5. Make marks on the tape at the 250 ml, 500 ml, 750 ml, and 1000 ml lines on the bag. These marks correspond to 15 minutes, 30 minutes, 45 minutes, and 60 minutes of time.
6. You will use the flow control dial to give the correct amount of fluid over time.

Mark a start line on the tape at the point where the fluid level and the tape meet. This is the start line. It should be at about the 1000 ml mark.

Place an Intravenous Catheter

Direct another dog handler to position and restrain the dog.

1. Have the dog restrained in a sitting position or in sternal recumbency (chest on the ground) that eliminates movement.
2. Muzzle the dog.
3. Have the handler restrain the dog's head by wrapping his arm, farthest from the dog, around the dog's neck and cradling the dog's head and neck in his elbow.
4. Have the handler wrap his fingers around the back of the dog's elbow and push the dog's front leg slightly forward to stabilize it.

Prepare the catheter site over one of the front leg veins, located on the front part of the leg about halfway between the elbow and the wrist.

1. Wet the area with 2–3 alcohol-soaked gauze sponges to remove gross dirt and smooth the hair. If alcohol is not available, use water.
2. Have the handler place the thumb over the vein and the heel of the hand under the dog's elbow and apply resistance in a forward direction. This will make it easier to see the vein as it swells with blood and prevents the dog from pulling the elbow backward.

Remove the catheter cover and hold the catheter in your dominant hand.

With your other hand, stabilize the leg and vein by placing your thumb directly alongside the vein and wrap your remaining fingers underneath and around the leg. It will appear that the leg is "cradled" in your hand.

Puncture the vein.

1. Pierce the skin with the catheter needle bevel (the angled tip of the catheter) facing up, at a 10 to 30 degree angle to the skin.
2. Advance the catheter to pierce the vein.

3. Confirm that you are in the vein by looking for a flash of blood at the hub of the catheter needle.
4. Decrease the angle of the catheter needle until it is almost parallel to the skin surface.
5. Advance the catheter needle approximately ¼ inch into the vein using a gentle forward motion.
6. Position the catheter.
7. Stabilize and hold the catheter needle hub with one hand.
8. Without moving the needle, advance the catheter into the vein as far possible with the other hand, only touching the hub of the catheter with the fingers.

Direct the dog handler to release the pressure on the vein, but continue to hold the elbow in place.

Remove the needle from the catheter by pulling it back and out while stabilizing the catheter to keep it in the vein.

- Blood will immediately flow out of the catheter if you have placed it correctly.
- Do not attempt to reinsert the needle into the catheter if blood is not flowing, as this could result in the catheter being sliced in half and the free end flowing into the heart. If the first attempt at catheterization is not successful, try placing a catheter in the other front leg.

Quickly attach the injection port adapter to the hub of the catheter to stop the flow of blood.

Secure the catheter to the leg with the tape.

1. Wrap the tape around the hub of the catheter then wrap the tape around the leg just below and under the cap, ensuring that the tape is not so tight as to restrict circulation.
2. Wrap a second piece of tape over the catheter injection port and around the leg.
3. Wrap a third piece of tape over the site where the catheter pierces the skin and around the leg to provide protection and additional security.
4. Flush the catheter.

5. Draw out 3 ml of sterile saline from the fluid bag using a 6 ml syringe and a sterile 22 gauge 1-inch needle.
6. Gently insert the syringe into the catheter injection port and inject the sterile saline in short, gentle spurts to flush the catheter.
7. The saline should flow smoothly and there should not be any swelling developing around the catheter under the skin.
8. If the saline does not inject easily or swelling is noted around the catheter, the catheter is not placed correctly and should be removed. A new catheter should be placed in the other front leg.
9. Roll self-adhesive conforming tape around the leg and catheter for further stability, ensuring that the catheter injection port is easily accessible.

Administer Intravenous Fluids

1. Remove the caps from both the line and catheter. **Be careful not to contaminate either end!**
2. Attach the line to the catheter. Make sure the fit is snug.
3. Slowly move the flow regulator back up to start the flow of fluids.
4. Note the time. Mark the start time, at the start line, on the tape marking the bag.
5. Observe the catheter insertion site for swelling. If there is swelling, the catheter is not correctly placed in the vein and may need to be replaced.
6. Set the drip rate at approximately 1–2 drops into the chamber per second. The drip rate is adjusted by rolling the flow regulator up or down.

Secure the administration set in place.

• Place a piece of tape around the dog's leg and the tubing, securing the tubing close to the catheter and tubing connection site. Create a courtesy tab at the end of the tape by folding the last ½ inch of tape onto itself.

• Form a loop with the IV tubing and place more tape around the dog's leg and loop of the tubing, securing the tubing to the dog's leg. Create a courtesy tab at the end of the tape.

Continue monitoring the catheter insertion site. Look for:

1. Swelling.
2. Blood.
3. Movement of the catheter.

Monitor the equipment. Look for:

1. Disconnects anywhere in the set up.
2. Kinks or blocks in the tubing.

Monitor the dog.

1. Make sure the dog doesn't bite or chew at the administration site. If the dog is conscious, you may need to keep the muzzle on to prevent chewing.
2. Monitor the dog's response to the fluid therapy. Be prepared to relay the information to the veterinary staff.
3. Monitor the fluid administration rate. Use the marks you created on the tape on the bag and a timepiece to ensure the administration rate is correct.
4. Adjust the drip rate by rolling the regulator up or down if the fluids are flowing too fast or too slow.

The goal of therapy is the controlled administration of 2000 mls of LRS over a 2-hour period.

1. Ensure that the MWD received 500 mls in the first ½ hour, 1000 mls by 1 hour, 1500 mls by 1½ hours, and 2000 mls by 2 hours.
2. Periodically reassess the MWD's vital signs.

Advise the kennel master and veterinary staff of the event and request further instructions. MEDEVAC any MWD that required intravenous fluid therapy.

Make a written record of the treatment, including the date, time, and actions taken.

PROVIDE FIRST AID TO MWD WITH GASTRIC DILATION–VOLVULUS (BLOAT)

When an MWD has severe abdominal distention, retching or nonproductive vomiting, and signs of pain (rapid and shallow breathing, anxiety, grunting, and weakness), and veterinary personnel are not available, you must initiate first aid for gastric dilatation–volvulus (GDV; bloat) without causing further harm to the dog.

Recognize the Three Hallmark Signs of GDV (Bloat)

There are varying degrees of abdominal distention from stomach filling with air, food, and fluid. NOTE: Many medical problems cause abdominal distention. It can be difficult to tell the difference between these and GDV/Bloat. However, if abdominal distention is present in addition to these other signs, assume GDV/Bloat is present and initiate first aid.

There can be nonproductive retching, attempted vomiting without result, retching a small amount of saliva, "dry heaves," and excessive salivating.

Signs of pain, if the dog is conscious:

- Grunting, especially when the stomach or abdomen is palpated.
- Anxiety, which is commonly noted as pacing, anxious stares, and inability to get comfortable when lying down.
- Panting.

Recognize Signs Associated with Shock

Recognize signs associated with shock such as weakness or collapsing; pale, gray, or blue mucous membranes; prolonged CRT; rapid heart rate; weak, rapid pulse; change in level of consciousness, from agitated to depressed to semiconscious, to unconscious.

Treat Shock

Administer intravenous fluids.

1. Place a 20-gauge intravenous catheter in a leg vein.
2. Give 1 liter of LRS each hour for 2 hours. Do not give more than 1 liter in an hour or more than 2 liters total.

If you don't have the necessary materials to give intravenous fluids, move on to decompress the stomach.

If oxygen is available, use a face mask or tubing and blow oxygen into the dog's nose or mouth at a flow rate of 5–10 liters per minute.

Decompress the Stomach to Relieve Gas Pressure

Lay the dog with its left side down and locate the insertion point.

- Feel the last rib on the right side of the dog.
- Find the point that is 2 finger-widths behind the last rib, halfway between the spine and the bottom border of the abdomen on the right side.

Wipe the area generously with gauze sponges soaked in alcohol.

Forcefully insert an 18-gauge intravenous catheter through the skin, abdominal wall, and stomach wall until you hit the hub at the end of the catheter.

- Leave the needle in the catheter.
- Ensure the abdominal wall and distended stomach are penetrated.
- The procedure is successful if gas or air comes through the trocar from the stomach. If no air or gas escapes, attempt the procedure one more time. If still unsuccessful, immediately transport the MWD to veterinary care and **do not** attempt the procedure a third time.

Attach a 3-way stopcock and 60-cc syringe to the catheter and gently aspirate air from the stomach using the syringe until you can no longer remove any air.

Remove the catheter once no further air can be removed, because leaving it inserted may cause trauma to internal organs.

Keep the catheter clean and repeat trocarization if abdominal distension recurs during evacuation.

Monitor the Dog and Your Treatment

Continue to monitor vital signs and note whether the dog is responding to treatment.

Continue to monitor the dog's abdominal area for further distention. Repeat stomach trocarization if significant abdominal distension develops.

Monitor the IV Catheter Insertion Site on the Leg
Look for:

- Swelling.
- Bleeding.
- Movement of the catheter. Once inserted, the catheter should not move.

Monitor the Intravenous Fluid Therapy Equipment
Look for:

- Disconnects anywhere in the set up.
- Kinks or blocks in the tubing.

Monitor Fluid Administration Rate
- Use the marks on the bag and a timepiece to ensure the administration rate is correct.
- Adjust the drip rate by rolling the regulator up or down if the fluids are flowing too fast or too slow.

Make a written record of the treatment, including the date, time, and actions taken.

PROVIDE FIRST AID TO MWD WITH AN OPEN CHEST WOUND

Open chest wounds can be caused by shrapnel, bullets, foreign objects (sticks, metal rods or pieces, etc.) or other objects that penetrate the chest wall. When an open chest wound is present, the lungs collapse, causing severe breathing problems and possibly death.

Recognize the Signs of an Open Chest Wound
- May be obvious or concealed by hair or blood.
- Probe any suspicious areas with a finger to see if the chest wall has been penetrated.
- Object impaled in the chest.
- Sucking or hissing sounds coming from a wound to the chest.

- Frothy blood coming from a wound on the chest.
- Labored or difficult breathing.
- Chest not rising as it should with a breath.
- Apparent pain with breathing.
- Signs of shock.

Provide First Aid for an Open Chest Wound by Immediately Covering the Wound as Described Below

1. Immediately place your hand directly over any chest wound to provide immediate protection.
2. If someone is available to help you, have that person place his or her hand over the wound.
3. Obtain a field dressing.
4. Check carefully for entry and exit wounds. If there is more than one wound, you will have to treat each wound separately.
5. Check for the presence of a penetrating object. If the penetrating object is still in the chest, DO NOT try to remove it.
   Treat for shock if indicated.

Apply an Air-Tight Cover to the Wound

1. Quickly cut a field dressing package all the way so that you have a flat piece of plastic. Place the paper-wrapped field dressing to the side. If a field dressing package is not available, try to find a similar item such as plastic sheet, cellophane, MRE wrapper, foil, or part of a poncho.
2. Squeeze some water-based lubricant on the plastic or other item, and spread it around.
3. Place the lubricant-covered, air-tight material directly over the wound to form a seal.
4. If there is more than one wound, use a separate plastic cover for each wound. If the penetrating object is still in the chest, do the best you can to form an airtight seal around the object.

Dress the Wound

1. Maintain pressure on the seal covering the wound.
2. With your free hand, shake the paper wrapper from the field dressing.

3. Place the dressing, white side down, directly over the seal covering the wound.
4. Secure the dressing by wrapping it around the dog's chest and tying the tails together.
5. Apply a bandage over the field dressing using non-adhesive conforming wrap that is wrapped around the entire chest. Apply this bandage with enough tension to keep the field dressing in place, but not so tightly that the dog has more difficulty breathing. **NOTE:** If the airtight seal is lost at any time during this process, start over. **An airtight seal must be maintained at all times.**

Notify the supporting veterinary personnel to request further instructions and contact the kennel master to advise him of the situation.

Make a written record of the treatment, including the date, time, and actions taken.

PROVIDE FIRST AID TO MWD WITH AN OPEN ABDOMINAL WOUND

Major organs such as the stomach, intestines, spleen, kidneys, urinary bladder, and liver are located in the abdominal cavity of the dog. These critical organs are susceptible to serious injury from ballistic wounds (e.g., gunshot, penetrating foreign body, shrapnel), blunt trauma (e.g., vehicular injury, falls), and blast injury (e.g., explosive devices and munitions). You must recognize and know how to provide first aid if your dog has an open abdominal wound to protect these organs and their blood supply.

- Obvious hole(s) in the abdominal area or an object impaled in the abdominal area.
- Lacerations or cuts and scrapes on the abdomen that appear to penetrate the abdominal wall.
- Exposed or protruding intestines or abdominal organs.
- Signs of shock.

Provide First Aid for an Open Abdominal Wound

Treat for shock if present and cover the wound

1. Immediately place your hand directly over the abdominal wound(s). If someone is available to help you, have that person place his or her hand over the wound.
2. Check for entry and exit wounds. If there is more than one wound, you will have to treat each wound separately.
3. Rinse the wound by pouring a liter of sterile LRS into the wound and over any exposed organs.
4. If there is a penetrating object still in the wound, do not remove it. Leave it where it is and work around it.
5. If internal organs have come out through the wound, do not try to push them back in. Wrap them in sterile gauze, dampen them with sterile LRS, and place them over the abdominal wound.
6. Apply a plastic cover to the wound.
7. Quickly cut a field dressing package all the way so that you have a flat piece of plastic. Place the paper-wrapped field dressing to the side. If a field dressing package is not available, use a similar item such as plastic sheet, cellophane, MRE wrapper, foil, or part of a poncho.
8. Squeeze some water-based lubricant on the plastic and spread it around.
9. Place the lubricant-covered plastic directly over the wound.

Dress the Wound

1. Maintain pressure on the plastic covering the wound.
2. With your free hand, shake second (paper) wrapper from the field dressing.
3. Place the dressing, white side down, directly over the plastic covering the wound.
4. Secure the dressing by wrapping it around the dog's abdominal area and tying the tails of the dressing.
5. Apply a bandage over the field dressing using non-adhesive conforming wrap that is wrapped around the entire chest. Apply this bandage with enough tension to keep the field dressing in place, but not so tightly that the dog has more difficulty breathing.

6. If the bandage becomes soiled by feces or urine, replace it immediately.
7. If your MWD is a male, try to avoid wrapping the prepuce in the bandage.

Notify the supporting veterinary personnel to request further instructions and contact the kennel master to advise him of the situation.

Make a written record of the treatment, including the date, time, and actions taken.

INDUCE VOMITING

An Emetic Is a Drug that Causes, or Induces, Vomiting

Emetics can be an important aspect in the treatment of orally ingested toxins. Determine if inducing vomiting is appropriate for your MWD.

Reasons to induce vomiting

- Ingestion of a toxic substance or training aid within the last 2 hours.
- Ingestion of a substance that is **not** corrosive or petroleum-based.

Reasons NOT to induce vomiting

- Ingestion of a toxic substance or training aid more than 2 hours ago.
- Your MWD is unconscious.
- Your MWD is unable to swallow.
- Ingestion of a corrosive or petroleum-based product, such as gasoline, oil, tar, grease, paint, solvents, paint strippers, paint thinners, nail polish or removers, hair spray, and batteries.

Induce Vomiting

1. Retrieve 1 of the 2 milligram apomorphine tablets from the aid bag.
2. Determine how much apomorphine is needed for your dog, which is based on body weight.
3. If the dog weighs less than 66 pounds, administer one-half tablet.
4. If the dog weighs more than >66 pounds, administer 1 full tablet.

5. Crush the tablet up and dissolve with a few drops of water. Place the correct amount of tablet in a 3-cc syringe.
6. Administer the apomorphine by gently pulling down on the lower eyelid to expose the conjunctiva. Place the entire amount of liquefied apomorphine directly onto the conjunctiva of the dog. Never administer more than 1 dose of apomorphine, even if vomiting does not occur. Repeated doses are unlikely to induce emesis and may cause apomorphine toxicity.

Take Action After Vomiting Has Started to Prevent Excessive Vomiting and to Monitor the Dog
- Apomorphine will cause intense vomiting in about 10–15 minutes.
- Excessive vomiting is dangerous to the dog.
- Fifteen minutes after vomiting started, thoroughly rinse the eyelid of unabsorbed apomorphine with at least one-half of a bottle of sterile eye rinse.
- Once the dog has vomited, DO NOT let the dog re-ingest the vomitus. If possible, save the vomitus for transport to the veterinary treatment facility for examination.
- Monitor the dog for any adverse effects from the apomorphine. Successful induction of emesis does not signal the end of appropriate monitoring or therapy. Adverse effects include: prolonged vomiting, excitement, restlessness, and respiratory depression.

Report Your Actions
Notify the supporting veterinary personnel to request further instructions and contact the kennel master to advise him of the situation.

Make a written record of the treatment, including the date, time, and actions taken.

APPLY A BANDAGE TO THE HEAD, NECK, OR TRUNK

As an MWD handler, you should be prepared to provide first aid for an injury to the head, neck, or trunk of your dog. Wounds need to be protected during evacuation. Bandaging incorrectly can cause further injury to your dog, so it is important to properly apply a bandage.

Advantages of Bandaging
- Speeds wound healing.
- Provides wound cleanliness.
- Controls bleeding.
- .Reduces swelling and bruising.
- Eliminates hollow spaces under the skin where fluid pockets and infection can develop.
- Immobilizes injured tissue (splinting).
- Minimizes scar tissue.
- Makes dog more comfortable.

Disadvantages of Bandaging
- May increase dog's discomfort.
- Dog may self-mutilate bandage and wound.
- Bacterial colonization of wound is greater.
- Ischemic injury (cuts off circulation).
- Damage to healing tissues.

General Principles of Bandage Application
1. Direct an assistant to position and restrain the dog so that the area to be bandaged is accessible.
2. Apply the first layer.
   - The first layer is in direct contact with the wound.
   - Cover the wound with a non-adherent dressing (e.g., Telfa® pad).
3. Apply the second layer.
   - The second layer holds the non-adherent dressing in place, adds wound protection, and absorbs fluid that comes from the wound.
   - Wrap 1–2 rolls of roll gauze around the affected area. Use firm even pressure when wrapping, but DO NOT wrap the gauze too tightly.
4. Apply the third layer.
   - The third layer provides support and additional protection.
   - Wrap conforming self-adhesive bandage (e.g., VetWrap) over the secondary layer without tension.

Bandaging the Head and Neck
1. Do not restrict breathing, swallowing, eating, or the eyes (unless covering an eye injury).
2. Ears may be left uncovered if not wounded.
3. If the ear is covered, mark the outline of the ear flap or write "ear flap up" or "ear flap down" on the bandage. This prevents anyone who removes the bandage from accidentally cutting the dog's ear.
4. Check tightness by placing 2 fingers under each side of the bandage. Your fingers should slip snugly under the bandage edges. If too tight, re-bandage with less tension.
5. Observe for difficulty swallowing, choking, or discomfort. If observed, re-bandage with less tension.

Bandaging the Thorax
1. The bandage must be wrapped in front of at least one of the front legs to prevent the bandage from slipping rearward.
2. Check tightness by placing 2 fingers under each end, and observe for difficulty breathing or discomfort. If observed, re-bandage with less tension.

Bandaging the Abdomen
1. Leave the prepuce exposed in male dogs to prevent urine soiling of the bandage.
2. Check tightness by placing 2 fingers under each end, and observe for difficulty in breathing or discomfort. If observed, re-bandage with less tension.
3. To prevent the bandage from slipping rearward, tape may need to be applied at the junction of the hair line and bandage material, along the leading edge of the bandage.

Monitor the Dog
1. Write the date and time the bandage was applied on the bandage.
2. Ensure that the bandage stays clean and dry. Replace if wet or soiled.
3. Observe for signs of pain or discomfort.
4. Observe for difficulty breathing. If a chest or abdominal bandage is too tight, it may interfere with respiration to the point of being life-threatening!

5. Check to see that the bandage has not slipped out of place
6. Sudden chewing at a bandage that has been previously well-tolerated is usually a sign the bandage is too tight.

Notify the supporting veterinary personnel to request further instructions and contact the kennel master to advise him of the situation.

Make a written record of the treatment, including the date, time, and actions taken.

PROVIDE FIRST AID FOR A PAD OR PAW INJURY

MWDs are very active on their feet, making their pads and paws susceptible to injury. Additionally, because their pads and paws are in contact with the ground, this area is open to all kinds of contaminants—especially if there is a wound. As a dog handler, it is an absolute necessity to learn how to give your dog first aid in case of a pad or paw injury.

Clean the Wound
1. Cleanse the wound with 250 ml of LRS. If LRS is not available, use sterile water, or, as a last resort, tap water.
2. Cut the syringe port off the fluid bag and squeeze the fluid directly on the wound.
3. Dry the foot very well before bandaging.

Apply a Bandage to the Paw
1. Position and restrain the dog so that the area to be bandaged is accessible.
2. Apply the first, primary, or contact layer by placing a non-adherent dressing (Telfa® pad) in direct contact with the wound.
3. Apply the second, secondary, or intermediate layer. Apply cotton cast padding starting at the end of the paw and working your way up the leg. Use firm pressure when wrapping and keep it smooth with no wrinkles. Overlap the previous roll by ½ each time and go all the way up to and include the accessory pad. NOTE: Never pull the gauze roll tight as you are applying it. Unless the middle two toes are injured, the bandage is applied such that these

two toes can be seen to check for swelling. The toe nails should be close together (almost touching) and parallel. If the toes are spreading apart, the foot is swelling and the bandage should be changed or removed immediately.

The Above-Noted Care for Your Dog's Paws Are Performed Under Ideal Non-threatening Conditions

The following can be followed when caring for your dogs pad or paw injuries in less desirable conditions and situations.

1. Cover the wound with some sort of bandage, preferably use 1" thickness of 4x4 gauze. If 4x4 gauze is not available try to find something sterile or clean (even a T-shirt can work in a pinch).
2. Apply firm pressure to the wound with the bandage between the wound and your fingers.
3. Using direct pressure to stop bleeding takes time. Avoid lifting up the bandage to look at the wound while waiting for the bleeding to stop. Looking under the bandage to look at the wound pulls off the blood clot that is forming. When the blood clot gets pulled off, bleeding takes longer to stop.
4. If the blood flow is strong enough to bleed through the bandage, apply an additional inch of 4x4 gauze and continue applying pressure until bleeding stops.
5. Apply the third outer layer. Wrap with rolled gauze (Kerlix) over the secondary layer without tension followed by elastic wrap (Vetrap) and/or adhesive tape (Elastikon) without tension. Start the Elastikon where the Vetrap ends. Wrap the tape around the paw or leg, ensuring that it is covering the top edge of the bandage and the dog's hair. Wrapping the tape too tightly will cause the paw to swell.

Monitor the Dog

- Ensure that the bandage stays clean and dry.
- Note if the dog's pain and/or discomfort level rises. If the bandage is too tight, it may interfere with circulation to the point of requiring an amputation.
- Check to see that the bandage has not slipped out of place.
- Sudden chewing at a bandage that has been previously well-tolerated is a sign of a problem.

Notify the supporting veterinary personnel to request further instructions and contact the kennel master to advise him of the situation.

Make a written record of the treatment, including the date, time, and actions taken.

APPLY A SPLINT OR SOFT PADDED BANDAGE TO A FRACTURE OF THE LIMB

The possibility always exists that your MWD may break its leg, and as a handler, it will be important for you to know how to prepare your dog for transport to the closest veterinary facility. In the instance of a broken bone, it will be important for you to know how to splint your dog's leg or apply a soft padded bandage to protect the leg during transport.

Recognize Fracture Type
Open fracture or compound fracture
- A broken bone has gone through the skin and is poking out.
- Because of the open skin, there is an elevated risk of infection.

Closed fracture or simple fracture
A closed fracture has no protruding bone. There are many other types of fractures, but open and closed are the basic two types with which you need to be familiar with.

Recognize Signs of a Fractured Limb
- The dog appears to be in pain.
- The dog will not put any weight on the affected limb.
- Swelling of the affected limb.
- The limb may look deformed, out of its normal shape.
- If the break is an open fracture, there will be a bone sticking out at the point of the break.
- The limb may be in an abnormal/awkward position.

Recognize Basic Dog Anatomy Requiring Splinting If Fractured. Only Fractures Below the Knee and Elbow Need to be Splinted

Limb anatomy
1. Front legs—note elbow and wrist.
   - Upper leg or humerus. The humerus is the big bone at the top of the front leg.
   - Lower leg has 2 bones, radius and ulna. These two bones are the smaller bones at the bottom of the leg. The ulna is the bigger bone toward the back of the leg. The ulna includes the elbow. The radius is the smaller bone in front of the ulna.
2. Rear legs—note stifle (knee) and hock (ankle)
   - Upper leg/thigh or femur. The femur is the big bone at the top of the rear leg.
   - Lower leg has 2 bones, fibula and tibia just below the knee. The tibia is the bigger bone in the front of the leg, and the fibula is the smaller bone toward the back of the leg.

Splinting limbs
- The bigger bones, the humerus and femur, are self splinting. If there's a lot of muscle around the bone, it will stabilize it naturally; if you try to add padding for a fracture above the knee or elbow, the extra weight of the padding on the end of the leg will act like a pendulum and make things worse.
- If broken, splint the bones below the knee and elbow (i.e., radius, ulna, fibula, or tibia).

Apply a Splint or Soft Padded Bandage to a Fractured Limb
Do not try to straighten a fractured limb, and handle a dog with a fractured limb with extreme care and caution.

Only apply a splint if the dog is unconscious and for temporary stability while transporting the dog to the veterinary facility. Only splint the bones below the knee or elbow (i.e., radius, ulna, tibia, or fibula).

Field expedient splint

1. Wrap a magazine or section of newspaper around the fracture.
2. Secure the magazine or newspaper around the fracture with heavy-duty tape such as masking tape, packing tape, duct tape, or "100 mile-an-hour" tape.

Two-tie splint

1. Place a sturdy stick or similar object on each side of the fractured limb.
2. Gently tie the sticks in place with one tie a few inches above the fracture and one tie a few inches below the fracture. The tie should be tight enough to maintain the leg in the position it is in.

Robert Jones™ bandage (soft padded bandage)

If there is bone sticking out of the fracture, cover it with something sterile (preferably) or clean.

1. Position and restrain the dog so that that area to be bandaged is accessible.
2. Apply 1-inch-wide tape stirrups to the inside and outside of the leg. Ensure that the ends of the tape extend about 4 to 6 inches below the foot. Fold the tape back on itself about ½ to 1 inch at the very end.
3. Place a tongue depressor between the two tapes where they extend beyond the paw. This will make it easier to handle the tape and to separate the tape later when it will be used in the bandage.
4. Apply roll cotton to the leg, starting from the paw end and working up the leg. Use a lot of padding. If you don't have roll cotton available, use towels, large first aid or medical dressings, or anything else you can think of to provide padding.
5. Apply conforming gauze to compress the cotton starting from the paw end and working up the leg. Ensure that you stabilize one joint above and below the fracture.
6. Remove the tape stirrups from the tongue depressor, twist ½ turn, and apply to the gauze on both sides of the leg.

7. Apply a layer of self-adhesive elastic wrap starting from the paw end and working up the leg.
8. Secure the elastic wrap with adhesive tape.
9. Test the bandage for firmness by thumping it with a finger. It should sound like a ripe watermelon.
10. Ensure the bandage is not too tight by slipping two fingers under the paw end of the bandage and loosen as necessary.
11. Write date and time on bandage.
12. Place an Elizabethan collar on the dog's neck if the dog is biting or licking the bandage.

Monitor the dog
- Ensure that the bandage stays clean and dry. If the bandage becomes wet, replace it immediately.
- Note if the dog's pain and/or discomfort level rises.

Make a written record of the treatment, including the date, time, and actions taken.

ADMINISTER ORAL MEDICATION

Dogs get sick or injured and require medication just as humans do. It is unrealistic to assume that there will always be a veterinary staff member available to give your dog its required medication; therefore, learning how to administer oral medications to your dog is a necessary skill.

Obtain Prescribed Medication
Verify the medication is what was prescribed by the veterinarian by checking the bottle for type of medication, strength, dosage, and so forth.

Ensure the medication is not expired by checking the expiration date on the bottle.

DO NOT give the medication if the medication has expired or is not the medication that was prescribed for the MWD without first consulting with veterinary personnel.

Prepare Medication for Administration

Tablets or capsules
1. Check the label of the bottle for dose.
2. Take out required dose.
3. Wrap the tablet or capsule in a meatball. It is easiest and safest to give a dog a tablet or capsule this way as most dogs will eat a meatball. If high-quality canned dog food is not available, wrap the tablet or capsule in a hot dog or a chunk of cheese.

Liquid medication
1. Check the label of the bottle for dose.
2. Draw up the required dose in a syringe or applicator provided with the medication.

Position and restrain the dog in the down or sitting position to administer the medication.

Administer the Medication
Tablets or capsules
Give the dog the prepared meatball. If the dog will not eat the meatball, then take the following steps:

1. Grasp the upper jaw with the palm of one hand resting on the dog's muzzle.
2. Lift and extend the dog's head.
3. Press the upper lips over the upper jaw teeth.
4. Apply gentle pressure directly behind the canine teeth.
5. Use the thumb and index finger. Do not cause harm to the dog by using too much force or pressure.
6. Pick up the tablet or capsule using the thumb and index finger or the index and middle finger of the free hand.
7. Open the dog's mouth by pushing downward on the lower jaw using the free fingers of the hand holding the tablet or capsule.
8. Place the tablet or capsule on the center, far back portion of the dog's tongue.
9. Hold the dog's mouth closed.
10. Ensure that the dog has swallowed the medication.

11. The meatball should be completely gone, as well as the tablet or capsule.

If manually administering the medication, massage the dog's throat with a gentle up and down motion until the dog swallows the tablet or capsule. If the dog does not swallow the capsule try the following:

1. Gently tap the nose or under the chin to startle the dog into swallowing the capsule.
2. Blow sharply into the dog's nostrils to cause the dog to swallow. This is not advised on a highly unruly or aggressive dog.
3. Place 1 to 2 drops of water on the dog's nose.
4. Open the dog's mouth to ensure that the tablet or capsule was swallowed.

Liquid medication
1. Tilt the dog's head so the nose and eyes are in one line (parallel to the floor).
2. Form a pocket by pulling out the dog's lower lip at the corner of the mouth.
3. Insert the syringe or applicator into the pocket using the free hand. Do not scrape the gums with the syringe.
4. Push the plunger of the syringe or applicator forward.
5. Administer the medication slowly in 3 to 5 ml increments.
6. Observe for swallowing while administering the medication. If the dog is wearing a muzzle, apply liquid oral medication the same as you would if dog is not wearing a muzzle. Pay extra attention to not scraping the gums with the syringe. Be aware that the dog may jerk its head suddenly. Ensure that if the dog jerks its head, your syringe and hand do not remove the muzzle.

Monitor the Dog Following Administration of Medication
Observe for possible side effects and inform the veterinary staff if any of the following symptoms appear:

- Drooling.
- Nausea.

- Vomiting.
- Depression.
- Diarrhea or change in stool.
- Coughing and gagging.
- Pawing at face.
- Any other reactions.

Make a written record of the treatment, including the date, time, and actions taken.

ADMINISTER EAR MEDICATION

As a dog handler, it may be up to you to administer oral ear medication in the event that your dog has an ear infection.

Obtain Prescribed Medication

Verify the medication is what was prescribed by the veterinarian by checking the bottle for type of medication, strength, dosage, and so forth. Otic medication is medication that is given in the ears. You may see "otic" rather than "ear" on a medication bottle. Remember that the two words mean the same thing.

Ensure the medication is not expired by checking the expiration date on the bottle.

DO NOT give the medication if the medication has expired or is not the medication that was prescribed for the MWD without first consulting with veterinary personnel.

Administer the Medication

1. Direct the dog handler to position and restrain the dog in the down or sitting position.
2. Expose the ear canal by holding the ear flap straight up.
3. Position the dispenser directly above the outside opening of the ear.
4. Do not touch any portion of the ear with the dispenser.
5. Administer the exact amount of the prescribed medication. It must go directly into the ear canal.

6. Gently massage the base of the ear.
7. Release the ear flap.
8. Repeat the procedure in the other ear, if applicable.

Make a written record of the treatment, including the date, time, and actions taken.

CLEAN THE EXTERNAL EAR CANALS

Clean the External Canals of an MWD
1. Position and restrain the dog in the down or sitting position.
2. Gently pull the ear flap straight up with one hand and hold it vertically. If the ear flap itself is extremely dirty, you will need to clean it.
3. Observe the ear canal for obvious deformities. Look for things such as lumps, sores, excessive debris, or offensive odors. DO NOT attempt to clean the ear if there are wounds, masses, or ulcers until you have consulted your local supporting veterinary personnel.
4. Apply ear cleanser so that the entire external ear canal is filled with cleaning solution. Ensure the dog does not shake the cleanser out of the ear until you have massaged the ear. Do not use any other liquid other than a labeled otic cleanser to clean the ears.
5. Massage the base of the ear to break up debris in the ear.
6. Allow the dog to shake its head.
7. Blot any excess cleanser using cotton balls or 4x4 gauze. Do not scrub or push debris down into the ear canal. NEVER use Q-tips or other similar devices to clean the ears and DO NOT attempt to clean the vertical or horizontal canals.
8. Ensure the ear is clean. If the ear still appears dirty, clean the ear again until no further debris is observed on the cotton balls or gauze.

Make a written record of the treatment, including the date, time, and actions taken and inform the veterinary staff of any abnormalities.

TRIM THE TOENAILS

Routine care of your MWD may require occasional nail trimming. Nails allowed to grow too long can lead to breakage, bleeding, and lameness. Long nails can continue to grow so much that they curl up and grow back into the paw. This situation must be avoided by trimming the toenails.

1. Muzzle the MWD and position and restrain the dog on its side to allow for easy access to the paws.
2. Hold the nail trimmers with the handle in the fingers of one hand and hold the dog's paw in the other hand.
3. Position the nail in the trimmers as close to the quick as possible with the cutting blade **away from the quick.** The quick is the central portion of the nail that is very sensitive and contains small blood vessels. The quick is also known as the nail vein. Cutting the quick causes pain, bleeding, and, in some cases, infection.
   • In light or white nails, the quick appears pink or darker than the nail. It is harder to see the quick in dark nails. If in doubt about the position of the quick, position only the very tip of the nail in the trimmers.
4. Cut the nail with the nail trimmers in one smooth motion.
5. Repeat the nail trimming process on every nail of each paw, as necessary.
6. Take immediate action to stop bleeding in the event you cut the quick.
   • Apply direct pressure to the bleeding nail tip with a 4x4 gauze.
   • If direct pressure does not stop the bleeding, apply a blood-clotting product in accordance with manufacturer's instructions.
   • If a blood-clotting product is not available, press the affected nail into a bar of mild soap.

EXPRESS THE ANAL SACS

Recognize Signs that Your Dog Has Impacted Anal Sacs

Scooting the hindquarters on the floor or ground, biting or licking the rectum or tail area, pain in the area of the tail or anus, and holding the tail in a funny position are all indicators your MWD has impacted anal sacs.

1. To express the anal sacs, position and restrain the dog in a standing position.
2. Ensure the dog is muzzled and ensure the dog is restrained in a standing position that prohibits movement.
3. Put on exam gloves and lubricate your index finger with water-based lubricant.
4. Insert the lubricated index finger into the rectum and locate the left anal sac. Anal sacs are typically found at the 4 and 8 o'clock positions. Each anal sac in an MWD is about the size and firmness of a small grape when filled with secretions.
5. Place 4x4 gauze or a folded paper towel over the rectum to catch secretions.
6. Gently squeeze the anal sac between the thumb and index finger, "milking" the fluid out toward the duct. It is possible to rupture an anal sac if you are not gentle. If you cannot manually express the anal gland, **DO NOT** keep trying and immediately notify veterinary personnel.
7. Repeat this procedure for the right anal sac.

Inspect anal sac fluid on the gauze or paper towel and note if there is blood or pus present.

Notify the veterinary staff immediately of any abnormalities such as blood or pus.

Make a written record of the treatment, including the date, time, and actions taken.

INITIATE MEDICAL EVACUATION

The Ability to Properly Initiate a MEDEVAC Request Is Imperative for Today's Dog Handlers

The following is the first aid and/or life-saving treatments that must be completed on your dog at your location prior to initiating a MEDEVAC or more commonly referred 9-line evac.

Immediately contact the senior military member present. This individual determines the need to request a medical evacuation and assigns precedence for such an evacuation.

Contact the dog's attending veterinary or supporting veterinary unit. The veterinarian or veterinary staff will ensure that the MWD is in stable condition and provide other guidance.

The attending veterinary unit generally coordinates with the MWD owning unit for medical evacuation of the dog and handler. As the dog's handler, you will travel with your dog when medical evacuation is required.

Attend to the dog's medical needs as required or instructed. Relay important information to appropriate personnel as required, provide restraint as necessary, and follow instructions for medical evacuation given to you by your superiors and veterinary services personnel.

Prior to the actual evacuation, it is imperative that you as the handler have a clear understanding of where you and your MWD are going to be evacuated to. This will help you ensure that veterinary personnel are at your destination.

Collect All Applicable Information Needed for a MEDEVAC (9 Line) Request

- Line 1—Determine the grid coordinates for the pickup site.
- Line 2—Obtain radio frequency, call sign, and suffix.
- Line 3—Obtain the number of patients and precedence.
- Line 4—Determine the type of special equipment required.
- Line 5—Determine the number and type (litter or ambulatory) of patients.
- Line 6—Determine the security of the pickup site.
- Line 7—Determine how the pickup site will be marked.
- Line 8—Determine patient nationality and status.
- Line 9—Obtain pickup site nuclear, biological, and chemical (NBC) contamination information, normally obtained from the senior person or medic (only included when contamination exists).

Record the Gathered MEDEVAC Information Using the Authorized Brevity Codes

Unless the MEDEVAC information is transmitted over secure communication systems, it must be encrypted. You must inform MEDEVAC personnel that patient is an MWD and also include handler and veterinary personnel in line 5.

Transmit a MEDEVAC Request
Contact the unit that controls the evacuation assets
- Make proper contact with the intended receiver.
- Use effective call sign and frequency assignments from the SOI.
- Announce clearly "I HAVE A MEDEVAC REQUEST"; wait one to three seconds for a response. If no response, repeat the statement.

Transmit the MEDEVAC information in the proper sequence
- State all line item numbers in clear text. The call sign and suffix (if needed) in line 2 may be transmitted in the clear.
- Line numbers 1 through 5 must always be transmitted during the initial contact with the evacuation unit. Lines 6 through 9 may be transmitted while the aircraft or vehicle is enroute.
- Follow the procedures provided in the explanation column of the MEDEVAC request format to transmit other required information.
- Pronounce letters and numbers according to appropriate radio/telephone procedures.
- Take no longer than 25 seconds to transmit.
- End the transmission by stating "OVER."
- Keep the radio on and listen for additional instructions or contact from the evacuation unit.

PROVIDE FIRST AID TO MWD FOR VOMITING OR DIARRHEA

Knowing Causes of Vomiting Will Assist You in Diagnosing Your Dog's Ailment
Digestive system problems, irritation or pain, drugs, toxic waste products (organ disease or failure), chemicals, inner ear problems, sudden change in diet, and other diseases are common reasons your dog may be vomiting.

Knowing Causes of Diarrhea Is Also Very Important
Irritation, infection, altered bacterial population, breakdown of the lining of the intestinal tract, and sudden change are some common reasons your dog may have diarrhea.

Recognize Signs of Vomiting and/or Diarrhea
The signs of vomiting and diarrhea are usually obvious and visible. If the animal does not vomit or have an episode of diarrhea right in front of you, you may find evidence of it in the dog's kennel.

Recognize Signs that Require Immediate First Aid or Attention from a Veterinarian
These signs include:

- Presence of fresh blood/blood clots in the stools or vomit.
- Black, tarry stools.
- Difficult or painful defecation. The dog will cry out in pain or stay hunched up trying to defecate for long periods of time.
- An abnormally swollen abdomen. This may indicate gastric dilatation or gastric dilatation–volvulus or some other abdominal disease or problem.
- The dog appears to be in pain or is depressed in addition to the vomiting or diarrhea.
- Increased body temperature.
- Profound dehydration.

Provide First Aid for Vomiting or Diarrhea
Vomiting
1. Take the vital signs and perform a physical examination of the dog.
2. Withhold all food for 24 hours after the last episode of vomiting.
3. Withhold all water for 12 hours after the last episode of vomiting.
4. Offer small amounts of water every thirty minutes after 12 hours.
5. After 24 hours, introduce small amounts of food several times a day over 2–3 days. The dog's full diet can be fed after 2-3 days.
6. Notify the closest veterinary staff if vomiting persists or there are any abnormalities on the physical exam.

Diarrhea
1. Take the vital signs and perform a physical examination of the dog.
2. Withhold all food for 24 hours after last episode of diarrhea. After 24 hours, introduce small amounts of food several times a day over 2–3 days. The dog's full diet can be fed after 2–3 days.

3. Allow the dog to drink as much water as it wishes.
4. Notify the closest veterinary staff if diarrhea persists or there are any abnormalities on the physical exam.

Monitor the dog for signs of continued vomiting or diarrhea.

Make a written record of the treatment, including the date, time, and actions taken.

PROVIDE FIRST AID TO MWD FOR EYE IRRITATION OR TRAUMA

Dogs May Have Mild Eye Irritation or Suffer Significant Eye Trauma
MWD handlers must recognize signs of eye irritation and trauma and know how to provide first aid for eye problems.

Indications and Signs of Eye Irritation
Tearing, rubbing of the eye or face, excessive squinting or holding the eyelids shut, excessive reddening of the white part of the eye, milky white discoloration of the outer clear part of the eye, and excessive discharge (greenish, yellowish, bloody) from the eye are all signs and indications of eye irritations.

Provide First Aid for Eye Irritation
If eye irritation is observed, immediately contact veterinary service personnel for guidance. If none can be reached, then administer first aid.

1. Have the dog muzzled, sitting, or lying upright on chest.
2. Ensure the dog's head is restrained enough to prevent sudden head movement.
3. Remove dry debris from around the eye using a dampened gauze sponge with sterile eye rinse. Gently stroke the eyelids and surrounding area. Do not rub the eye directly. Allow the dog to close the eye during this procedure.
4. Flush moist secretions, dirt, blood, and discharge from the eye with approximately one-fourth of a bottle of sterile eye rinse.
5. Use the heel of the hand holding the eye flush to pull the upper eyelid open.

6. Use the thumb of the hand holding the jaw to pull the lower eyelid downward, exposing the inside part of the lower eyelid.

Apply sterile antibiotic eye ointment to the affected eye

1. If the ointment has been used before, squirt a small amount onto a gauze sponge to remove possible contaminants.
2. Use the heel of the hand holding the medication to pull the upper eyelid open.
3. Use the thumb of the hand holding the jaw to pull the lower eyelid downward, exposing the inside part of the lower eyelid.
4. Administer the medication.
5. To prevent contamination or injury, do not allow the tip of the tube to touch the eye or any other surface.
6. Squeeze the ointment tube to lay a single ribbon of ointment ¼ to ½ inch long) directly on the inside part of the lower eyelid.
7. Allow the dog to blink. If the dog does not blink, gently move the eyelids together to spread the ointment over the eye.

Recognize the Signs of Eyeball Trauma
- Blood or bruising in or around the eyeball.
- Cuts or wounds to the eyeball or surrounding skin.
- Foreign object in or around the eyeball.
- Displacement of the eyeball from the socket.

Provide First Aid for Eyeball Trauma
1. Clean and rinse the eye as noted above.
2. DO NOT pull on the eye if it is displaced. DO NOT try to put the eyeball back in the eye socket if it is displaced. DO NOT try to remove any object that seems to be embedded in the eyeball. These actions may cause more damage to the eye.
3. Gently coat the eyeball with the entire contents of one tube of sterile antibiotic eye ointment.
4. Cover the eyeball with 5-8 gauze sponges that have been moistened with sterile eye rinse.
5. Apply a bandage over the gauze to protect the eye.
   - When bandaging the head and neck, do not restrict breathing, swallowing, or eating.

- Ears may be left uncovered if not wounded. If the ear is covered, mark the outline of the ear flap or write "ear flap up" or "ear flap down" on the bandage to prevent anyone who removes the bandage from accidentally cutting the ear.

- Check tightness; you should be able to slide two fingers under each edge of the bandage. If too tight, re-bandage with less tension.

- Observe for difficulty swallowing and for choking or discomfort. If observed, re-bandage with less tension.

Inform the kennel master of the situation and contact supporting veterinary personnel to request further instructions.

Make a written record of the treatment, including the date, time, and actions taken.

ADMINISTER AN ANALGESIC INJECTION

Unfortunately, your dog may experience pain, whether due to a traumatic accident or some other medical reason. As the individual closest to your dog, it will be up to you to recognize when your dog needs pain-relieving medication and give it to your dog without causing further harm.

Recognize the Signs of Moderate to Severe Pain

Pain in an MWD may be caused by trauma, injury, or significant disease. Any MWD with these signs may need to be treated for pain. Contact your nearest veterinary support personnel FIRST. If no veterinary personnel can be reached, then you must treat for pain by following the specific steps described in this task. These are signs of moderate to severe pain:

- Vocalizing (barking, howling, whimpering, crying out, groaning) continuously or intermittently, especially when touched or moved, or when the dog is trying to move.

- Change in behavior, usually becoming more aggressive than normal or becoming depressed, anxious, nervous, or obviously uncomfortable.

- Uncomfortable appearance, with the dog acting restless or avoiding certain positions.

- Not putting weight on a limb or affected part of the body, or vocalizing when attempting to bear weight or use a body part.

Administer an Analgesic Injection

Assemble the needle and syringe.

1. Select a sterile 22 gauge x ⅝-inch needle and a sterile 3 ml syringe from the first aid kit.
2. Test the plunger for smooth, easy movement and a tight seal by pulling the plunger back and forth.
3. Connect the needle and syringe by screwing them together. DO NOT contaminate the needle or syringe—hold both only by the protective cap or covers, and do not touch the bare needle or tip of the syringe at any time. Once the needle is screwed into place, remove the outer package. The cap should remain over the needle.
4. Draw the correct amount of pain medication (morphine, 15 mg/ml) into the syringe.
   a. Inspect the medication vial to ensure it is the proper medication and strength and that it is not expired.
   b. Check the dose chart and determine the volume of pain medication to give based on the MWD's known or estimated body weight.
   c. Wipe the top of the medication vial with a gauze sponge soaked in alcohol.
   d. Remove the needle cap and insert the needle into the medication vial and withdraw slightly more medication than necessary, while holding the needle, syringe, and medication bottle vertically with the needle pointing up.
   e. Work the air bubbles out of the syringe.

Table 3. Analgesic Injection Dosage Chart

Body weight (in pounds)	Volume of morphine to give (in milliliters)
30 to 40	0.4
40 to 50	0.5
50 to 60	0.6
60 to 70	0.7
70 to 80	0.8
80 to 90	0.9
90 to 100	1.0

NOTE: A copy of this dose chart **needs** to be included in all medical kits. *It is mandatory* that all handlers know the weight of their MWDs and the appropriate dose.

1. Hold the needle, syringe, and medication bottle vertically with the needle pointing up.
2. Tap the syringe with a finger to work air bubbles up toward the needle hub.
3. Push the plunger to expel the bubbles back into the medicine vial.
   f. Push the plunger of the syringe forward until the **prescribed** amount of medication remains in the syringe.
   g. Remove the syringe/needle from the vial.
   h. Replace the needle cap.
5. As an alternative, a 10 mg morphine auto injector can be used if you are issued one of these for your canine.

Inject the Medication

- Have the dog restrained in a standing position that eliminates movement. If the dog cannot stand, it is okay to administer the injection with the dog lying down on its side.
- Ensure the dog is muzzled.
- Ensure the injection site is accessible. Intramuscular injections are administered in the muscle mass behind the femur bone in the upper, rear leg, avoiding the large nerve (sciatic nerve). NOTE: The sciatic nerve is a large nerve that runs parallel to the femur. If you do touch the sciatic nerve you will cause pain and possible paralysis.

Prepare the intramuscular injection site

Isolate the muscle receiving the injection. There are two acceptable ways to do this.

1. Wrap your thumb around the front portion of the thigh with your hand on the inside of the leg from the front to the rear.
   - Use your fingers to push out the muscle from the inside to the outside.
   - With the tip of your thumb, cover the area where the sciatic nerve lies on the outside to protect it from the injection.

2. Wrap your hand around the rear portion of the thigh with your hand on the inside of the leg from the rear to the front.
   - Use your hand to push the muscle up.
   - Use your thumb to block off the sciatic nerve.

Apply alcohol to the injection site. With the free hand, apply alcohol using a gauze sponge soaked with alcohol to wet the skin. Gently rub the skin to remove dirt and skin oil.

Inject the medication
1. Insert the needle into the muscle quickly and firmly.
2. Insert the needle at a 90° angle to the skin.
3. Insert the needle ½ inch to ¾ inch.
4. Pull back on the syringe plunger to ensure the needle is not in a blood vessel.
   a. Pull back on the plunger of the syringe.
   b. If the needle is in a blood vessel, blood will flash into the hub and syringe.
   c. If a blood flash appears, immediately remove the needle. DO NOT INJECT!
5. Press the plunger steadily forward while holding the syringe barrel steady.
6. When finished injecting the medication, withdraw the needle.
7. The drug's effects are usually felt within 15 to 30 minutes and may last up to 8 hours. These effects may include but are not limited to pain relief; sedation; apathy; reduced tension, anxiety, and/or aggression.

Dispose of the needle and syringe in accordance with local standard operating procedures (SOP).

How to Give the Auto Injector (Same Site as the Previous)
The U.S. military utilizes auto injectors; the dosage is usually 1 auto injector equals 10 mg. Auto injectors come in multiple strengths. Make certain you always double-check the dosage on the auto injectors.

- The dog handler will remove the safety cap; press the colored end into the proper injection location and press firmly. This will depress the black firing plunger.
- This action causes a release of gas within the injector that drives the hypodermic needle through the protective cap and about an inch into the thigh muscle. It also pushes 1.0 ml of the fluid, containing 10 mg of morphine, through the needle and into the muscle tissue.

Document the injection and the time administered.

Monitor the Dog Following Administration of Medication

Observe for possible side effects and inform the veterinary staff if any of the following symptoms appear: nausea/drooling, vomiting, depression, diarrhea or change in stool, change in breathing pattern, skin rash, swelling at the injection site, and any other reactions.

Immediately contact the closest veterinary staff to request further instructions.

Make a written record of the treatment, including the date, time and actions taken.

1. Note medication and dosage given to the dog and reason medication was administered.
2. Do not administer additional analgesic medication to your dog unless specifically directed to do so by your supporting veterinary personnel. Additional analgesic medication may cause life-threatening decreases in blood pressure or cause the dog to stop breathing.

PROVIDE FIRST AID FOR A BURN

It is important to determine what type of burn and the severity of the burn the dog has prior to initiating treatment.

There are many types of burns, but the following two are the typical burns affecting MWDs.

- Thermal burns are from radiation or some other type of heat source.
- Chemical burns are from a caustic chemical.

Determine Severity of the Burn

Burns are classified in degrees. The more layers of skin affected by the burn, the higher the degree classification. The higher the degree classification, the more serious the burn injury.

- First degree (superficial) burn—redness and pain, similar to a sunburn.
- Second degree (partial thickness) burn—red or mottled appearance, swelling, extreme hypersensitivity leading to pain. The area may appear wet and weeping due to fluid swelling up below the skin.
- Third degree (full thickness) burn—dark and leathery appearance. Skin surface is dry. There is no pain as the nerve endings are destroyed. If any hair remains, it will pull out easily.

Provide First Aid for a Burn

1. Perform a survey of the dog and take immediate action, if required.
2. Take the patient's vital signs.
3. Provide first aid for shock, if indicated from the vital signs.
4. Initiate appropriate burn treatment based on the type of burn.

Thermal burn

1. Apply cold compresses made of sterile water or saline-soaked towels, lap sponges, or gauze for a minimum of 30 minutes.
2. Submerge the dog in a cold water bath if the burn covers a large surface area.
3. DO NOT use ice, ice water, or iced saline or attempt to clip hair from the burned skin.
4. Carefully flush the burned surface with chlorhexidine solution (i.e., Nolvasan®) to remove surface debris.
5. Thoroughly flush with saline solution after the chlorhexidine solution flush.
6. Apply a thin coating of silver sulfadiazine ointment. Make sure you are wearing sterile gloves, not exam gloves, or use sterile applicators.
7. Apply sterile nonstick pads (i.e., Telfa® pad) to all wounds and loosely bandage the affected area with dry, sterile bandage material.

Chemical burn

1. Flush the burned area with sterile water or saline thoroughly for a minimum of 30 minutes to remove residual chemical.
2. DO NOT attempt to clip hair from the burned skin.
3. Carefully flush the burned surface with chlorhexidine solution (i.e., Nolvasan®) to remove surface debris.
4. Thoroughly flush with saline solution after the chlorhexidine solution flush.
5. Apply a thin coating of silver sulfadiazine ointment. Make sure you are wearing sterile gloves, not exam gloves, or use sterile applicators.
6. Apply sterile nonstick pads (i.e., Telfa® pad) to all wounds and loosely bandage the affected area with dry, sterile bandage material.

Immediately contact your closest supporting veterinary staff and request further instructions. Also, contact the kennel master immediately and inform him of the situation.

Continue to monitor the dog's vital signs. Be aware that pulmonary edema from smoke inhalation can occur after 24 hours, even if the dog shows no outward signs of being burned.

Make a written record of the treatment, including the date, time, and actions taken.

PROVIDE FIRST AID FOR A COLD INJURY

Identify Cold Injury Risk Factors

Always check your dog's records for a history of exposure to extreme cold. Several risk factors make a dog more susceptible to cold injury including but not limited to, inadequate acclimation, previous cold injuries, fatigue, inactivity, geographic origin, medications, poor nutrition, and dehydration.

Identify Individual Protective Measures Used to Avoid Cold Injuries

The military mission and situation may be such that you will not be able to incorporate all of these measures. In whatever situation you find yourself, take the best possible care of you and your dog. Avoid cold exposure, if possible. If mission dictates being out in the cold, take the following measures:

1. Avoid wind exposure. The wind chill factor should be considered if there is any wind. Wind combined with cold temperatures creates a wind chill factor, which is actually colder than the temperature on the thermometer.
2. Avoid contact with frozen objects.
3. Avoid fatigue. Use work/rest cycles to allow for rewarming.
4. Allow the dog time to acclimate to the environment and workload gradually. Partial acclimatization takes approximately 4–5 days, whereas full acclimatization takes 7–14 days.
5. Gradually increase physical activity.
6. If your dog has had a previous cold injury or if the dog is on medication, discuss the situation with your supporting veterinary staff.
7. Address nutritional needs by providing hot fluids and/or increasing the dog's caloric intake. Always discuss food or water changes with your supporting veterinary staff beforehand.

Describe the Two Types of Cold Injuries and Their Clinical Signs

Hypothermia—below normal body temperature

Preventing hypothermia must take place when the dog's temperature is still above 98°. Clinical signs of hypothermia include rectal temperature below 95° (dog may also be cold to the touch), dog is unconscious or appears to be "out of it," slow breathing, slow pulse rate, shivering may be present or may have stopped, weakness, and shock.

Frostbite—frozen body tissues

Signs of frostbite include dog's tissues are very cold to the touch; skin appears gray, white, or waxy instead of pink; and blistering occurs in more advanced cases. The tips of the dog's ears, scrotum, tail, lower legs, paws, and toes are the most common areas effected by frostbite. Frostbite may be most noticeable at the toenail beds and/or the edge of the ear flaps.

Provide First Aid for a Cold Injury

Cold injuries are progressive meaning that they get worse as time passes. Take immediate action as soon as you recognize the onset of a cold injury. The dog must be warmed as quickly as possible.

Treating for hypothermia
1. Warm the dog by wrapping the dog in blankets or towels if available, or place the dog on a circulating warm water heating pad, ensuring a water temperature of 85° to 103°F.
2. Wrap the water pad in towels or a blanket and cover the dog and pad with a blanket.
3. If a water pad is not readily available, old expired IV bags can be heated and used to warm the dog. Wrap the bag in a towel to prevent burns.
4. Ensure you use a heater to warm the room and monitor the dog's rectal temperature every 15 minutes.

Treating for frostbite
1. Warm the affected area by placing the injured part(s) of the dog in warm water (85° to 103°F) for 15 to 20 minutes or by applying warm wet towels to the affected area for 15 to 20 minutes, changing the towels every 5 minutes.
2. Gently pat dry the injured area. DO NOT rub.
3. To prevent self-trauma to the affected area or if it appears that the dog will not leave the area alone, apply an Elizabethan collar if one is available or muzzle the dog.

Inform the kennel master of the situation, and immediately contact the closest veterinary staff and request further instructions.

Make a written record of the treatment, including the date, time, and actions taken.

ADMINISTER ACTIVATED CHARCOAL

Activated charcoal is administered to an MWD to induce vomiting should your dog eat a toxic substance. Ensure when administering the activated charcoal that the dog is conscious and is able to swallow.

Activated charcoal would **NOT** be administered to an MWD if 1 hour has passed since the dog ingested the toxic substance.

Do NOT Induce Vomiting
DO NOT induce vomiting if the MWD ate a corrosive or petroleum-based product such as, but not limited to, gasoline, oil, tar, grease, paint,

solvents, paint strippers, paint thinners, nail polish or removers, hair spray, or batteries.

Administer Activated Charcoal by Mouth
DO NOT attempt to administer activated charcoal orally if the dog is not conscious.

- Determine how much activated charcoal your dog needs based on its body weight. For dogs weighing between 40–60 pounds, give 2 bottles of Toxiban™ (240 ml/bottle) orally. For dogs weighing more than 60 pounds, give 3 bottles of Toxiban™ (240 ml/bottle) orally.
- You should have 3 bottles of Toxiban™ in your aid bag. Retrieve the proper amount of bottles and use a 60 cc syringe to measure the required amount.
- Administer the required amount of Toxiban™ in one of two ways, as an oral slurry or mixed with food.

Administer Toxiban™ as an oral medication (slurry)
1. Tilt the dog's head up so the nose is pointing mostly toward the sky.
2. Form a pocket by pulling out the dog's lower lip at the corner of the mouth.
3. Insert the syringe gently into the pouch using your free hand. Do not scrape the gums with the syringe.
4. Push the plunger forward slowly to squirt the medicine onto the dog's cheek pouch.
5. Administer the medication in 3 to 5 ml increments.
6. Watch for swallowing between squirts and give your dog enough time to swallow each squirt of charcoal before giving your dog more.

Administer Toxiban™ mixed with a high-quality canned dog food
1. Mix a small amount of Toxiban™ (one-quarter of the bottle) with an equal amount of high-quality canned dog food (if available). Give it to the dog. If the dog consumes it, administer the rest of the dosage mixed with an equal amount of dog food.
2. If the dog does not consume the Toxiban™ mixed with dog food, you will have to administer the medication as an oral slurry.

Contact veterinary support for further instructions, and inform the kennel master of the situation.

Make a written record of the treatment, including the date, time, and actions taken.

TREAT AN MWD FOR TRAINING AID TOXICITY

MWDs will frequently come in contact with explosive and narcotic training aids when conducting training. It is imperative that handlers be able to recognize and provide first aid to an MWD who is displaying signs and symptoms of training aid toxicity.

Symptoms of Training Aid Toxicity for a Nitrate and Nitro-Based Explosive Training Aid

Symptoms of training aid toxicity for a nitrate and nitro-based explosive training aid that has been ingested are salivation, dizziness, stumbling, nausea, convulsions, dark brown mucous membranes, cyanosis, and death.

Symptoms of Training Aid Toxicity for Narcotic Training Aids

- If a marijuana/hashish training aid is ingested, MWDs will show signs of confusion, hallucinations, dizziness, nausea, and having breathing problems.
- If a heroin training aid is ingested, MWDs will have pinpointed pupils, slow heart rate, and breathing problems, and possibly go into a coma.
- If cocaine or amphetamine is ingested, MWDs will have dilated pupils; show signs of restlessness and aggression; hallucinate; and have a rapid heart rate and convulsions.

Treat an MWD for Training Aid Toxicity

If you witness your dog ingest a training aid, or the dog is displaying symptoms that it has recently ingested a training aid, take immediate action by inducing vomiting. Try to keep the dog calm by speaking reassuringly to and around it.

Immediately contact the closest veterinary staff to request further instructions.

Make a written record of the treatment, including the date, time, and actions taken.

PERFORM NUCLEAR, BIOLOGICAL, AND CHEMICAL DECONTAMINATION

The decision to decontaminate and treat an MWD or other military animal will be based on local SOP, theater restrictions, and other factors. These decisions will be directed by the theater commander, senior medical commander in theater, or senior veterinary commander in theater. The steps outlined in this task are generic in nature and are based on current available doctrine. Modifications to these steps may be necessary based on numerous factors, and the commander must direct the specific steps to be followed for a given situation.

Decontaminate for Nuclear Fallout
Remove radioactive particles from the hair coat and skin by brushing and bathing the animal in soap and water.

Decontaminate for Biological Agents
Wash the animal with soap and water, or follow command directives or policies for specific agents.

Decontaminate for Irritant Agents
These agents have little effect on animals. Flush the eyes with copious amounts of water or saline if liquid or solid agents come in contact with the eyes.

Decontaminate for Nerve Agent
Decontaminate hair and skin using the M291 skin decontamination kit.

1. Protect the eyes by applying a generous amount of ophthalmic ointment or similar nonmedical ointment (e.g., petroleum jelly).
2. Rinse the animal thoroughly to remove the decontamination solution.
3. Bathe the animal with warm, soapy water, and rinse thoroughly.
4. Decontaminate the eyes by irrigating with copious amounts of water or saline until all agent has been removed. Avoid using

any components of the M291 skin decontaminating kit in the eyes.
5. Decontaminate collars, leashes, muzzles, cages, bowls, and other items using M291 or M295 decontamination kits prior to putting them back on the MWD or using them.

Decontaminate for White Phosphorus
1. Immediately cover the affected area with water by submersion or with water-soaked bandaging material.
2. As quickly as possible, bathe the affected part in a bicarbonate solution to neutralize the phosphoric acid.
3. Remove remaining white phosphorous fragments (these are visible in dark surroundings as luminescent spots).
4. Treat the animal for thermal burns once all phosphorus has been removed.

Decontaminate for Blood Agents
Use the M291 skin decontamination kit.

Decontaminate for Blister Agents
Decontamination for blister agents (mustard, nitrogen mustard agent, or arsenical blister agents) should be carried out within 1–2 minutes after exposure using the M291 skin decontamination kit.

Decontaminate for Incapacitating Agents (BZ Type)
Decontaminate for incapacitating agents (BZ Type) by washing the hair coat and skin with warm, soapy water.

Dispose of wastes in accordance with local SOP.

PERFORM LIFE-SAVING THERAPY FOR ORGANOPHOSPHATE OR CARBAMATE POISONING

If your MWD is exposed to chemicals and is now acting unusual, perform a primary survey, take the dog's vital signs, and take the following measurements and record them:

- Body temperature.
- Pulse rate.
- Respiration rate.
- CRT.
- Mucous membrane color.

Signs of Organophosphate/Carbamate (Pesticide) Toxicity

Excessive salivation, drooling, muscle twitching (usually begins with the face and progresses over the entire body and becomes much more severe as attempts at walking becomes stiff and jerky), difficulties breathing, convulsions, urination, tearing, defecation/diarrhea, and death due to respiratory failure are all signs of organophosphate/carbamate (pesticide) toxicity.

Gather additional history from the person who has been with the dog for the past 12 hours.

- Were there unusual odors in the kennel or work environment?
- Was the kennel area sprayed or fogged for insects recently?
- Was the dog exposed to unusual dust or powder?
- Was anything applied to the dog externally (i.e., shampoo, ointment, drops, etc.)?
- Is there any other situation that you can recollect that might have involved a chemical exposure?
- Smell the hair coat for the odor of insecticides or to determine if the dog has any unusual odor.
- While wearing gloves, use your hand to part the fur in several places on the dog's body and smell for chemicals.
- Inspect the dog's feet and legs and smell them for chemicals. Look at the fur to see if there are any dark, wet, or damp spots on the dog that might indicate the dog got something on it.

Initiate life-saving therapy for organophosphate/carbamate toxicity

Administer atropine to your MWD intramuscularly using the dosing chart below (ensure that a copy of this chart is stored with the medication):

Table 4. Atropine Dosage Chart

Weight of Dog	Atropine Dose (15mg/ml)
50 lbs	.45 ml
60 lbs	.55 ml
70 lbs	.65 ml
80 lbs	.70 ml
90 lbs	.80 ml
100 lbs	.90 ml

NOTE: If after administering atropine, the dog's heart rate increases or the pupil's dilate, DO NOT administer more atropine, as it is unlikely to be organophosphate or carbamate toxicity.

1. Watch for improvement of the patient or disappearance of signs within 3 to 10 minutes.
2. Place an IV catheter.
3. Initiate an IV infusion.
4. Continue to monitor the MWD by:
   a. Take the dog's vital signs.
   b. Watch for muscular twitching to intensify.
   c. Watch for improved breathing and breathing pattern.
   d. Monitor mucous membrane color and how moist the gums and tongue are.
   e. Check eyes and note if pupils are dilated or pinpoint.
   f. Check CRT.

Lower Elevated Body Temperature If It Is Over 104°F
Drench the dog with cool water, wrapping the dog in wet towels and by placing a fan near the dog to keep it cool.

Reduce Continued Exposure to the Toxic Agent
- If the exposure was through the skin or fur, bathe the dog with a mild shampoo (NOT a flea/tick shampoo) and rinse thoroughly with warm water.
- If the exposure was by mouth, administer activated charcoal.

Repeat Atropine Administration If Clinical Signs Return or Intensify
Use the same dose from your dose card given in the muscle every 4–6 hours.

Chemical Exposure Source
If you haven't determined it yet, continue to try to find the source of the chemical exposure.

Immediately contact the closest veterinary staff to request further instructions.

Evacuate the MWD to the nearest veterinary facility if the situation dictates.

Make a written record of the treatment, including the date, time, and actions taken.

SYMPTOMS AND CONTROL MEASURES OF DISEASES AND PARASITIC INFECTIONS

Animal parasites survive by feeding from the dog's body and are harmful to the animal's health.

Hookworms
The most harmful intestinal parasite, hookworms live primarily in the small intestine and are typically 1/5–2/5 inches in length.

- *Symptoms:* Pale mouth and eye membranes, loose stools containing blood and weight loss.
- *Control:* Primarily by feeding rations with a chemical to prevent worms from completing a life cycle, and by keeping the dog's living area sanitary and free of stools.

Roundworms
An internal parasite that robs the infected animal of vital nutrients while living in the intestines, roundworms can be up to 6 inches in length.

- *Symptoms:* Diarrhea, vomiting, loss of weight, and coughing. The worms (spaghetti like) may be noticed in the stool or vomitus.
- *Control:* Done by treating the infected animal and kennel sanitation.

Whip Worms

Whip worms can be microscopic to 2 inches in length.

- *Symptoms:* Diarrhea, loss of weight, and paleness of mouth and eye membranes.
- *Control:* Same as for roundworms.

Tapeworms

Tapeworms are long, flat, ribbon-like, and segmented. They infect intestines and are noticed in the dog's stool as tiny whitish objects ¼ inch in length.

- *Symptoms:* Not very noticeable but may include diarrhea (often with blood or mucus), loss of weight, and decreased appetite.
- *Control:* Treatment of infected animals, good sanitation, and control of fleas.

Heart Worms

Thread-like parasites, heart worms are 6 to 11 inches long, are found in the heart and lungs, and interfere with the dog's cardiovascular functions.

- *Symptoms:* Coughing, loss of weight, difficult breathing, and loss of energy or stamina. The veterinarian can diagnose the disease with a blood test.
- *Control:* Feeding rations with a chemical that terminates the life cycle of a heart worm and controlling mosquitoes in the area.

Ticks

Common in many parts of the world, ticks attach themselves to the skin and suck the animal's blood, and may transmit disease.

- *Symptoms:* Small bumps on the skin. Take extreme care in their removal, as they may carry diseases harmful to humans.
   a. Grasp the tick as close to the skin as possible (a pair of tweezers is recommended).
   b. Pull slowly and gently until the tick is removed.
   c. Examine the tick to make sure you removed the head and

body from the dog, and did not separate the tick's head from the tick's body, leaving it attached to the dog.

- *Control:* Spraying the kennel runs and kennel areas with insecticide.

Fleas

Fleas torment the dog and spread disease and tapeworms.

- *Symptoms:* Found on the dog's skin and crawling in the hair.
- *Control:* Individual treatment and kennel sanitation.

Lice

- *Symptoms:* Small white or gray crescent-shaped objects fastened to the dog's hair.
- *Control:* Treatment of infected animals.

Mites

There are two types of mites: ear and mange.

- *Symptoms:* Ear mites—the dog will shake and/or scratch its head, and a brown (often dry) discharge from the ear(s) may occur. Mange mites—the dog may experience hair loss, scabbing/crusting skin lesions, and/or skin infections.
- *Control:* Treatment of infected animals by the attending veterinarian.

CANINE INFECTIONS AND DISEASES

Microscopic organisms cause contagious diseases transmitted from animal to animal. Zoonotic diseases are contagious diseases transmittable from animal to man. The following diseases, symptoms, and control measures apply:

Canine Distemper

Widely spread, canine distemper is highly contagious and usually fatal.

- *Symptoms:* Elevated temperature, loss of appetite, depression, loss of weight and energy, diarrhea, vomiting, coughing, thick discharge from eyes and nose, muscle stiffness, and convulsion.
- *Control:* Proper sanitation and immunization.

Canine Hepatitis
Found mostly in young dogs, canine hepatitis is spread through urine of infected animals.

- *Symptoms:* Same as canine distemper.
- *Control:* Immunization and sanitation.

Leptospirosis
Known commonly as "lepto," leptospirosis is caused by a microorganism called a spirochete and transmittable to man.

- *Symptoms:* Same as canine distemper.
- *Control:* Immunization, rodent control, and thorough cleanup after treating infected animals.

Rabies
Rabies is a disease that, like Lepto, is transmittable to man, but the transmission is through the saliva of an animal bite.

- *Symptoms:* May include sudden change in temperament or attitude, extreme excitement, difficulty in swallowing water or food, a blank expression, slackened jaw, excessive drooling from the mouth, paralysis, coma, and eventually, death.
- *Control:* Vaccination is needed. Handlers must prevent contact between their dogs and wild or stray animals. Report contact resulting in bites or scratches to the veterinarian. Capture the biting animal and hold for observation until released by the veterinarian. Use extreme caution during the capture to prevent bites to personnel.

Other Contagious Infections and Diseases
Vaccine cannot treat upper respiratory infection, pneumonia, and gastroenteritis.

- *Symptoms:* High temperature, loss of appetite, loss of energy, vomiting, diarrhea, and coughing.
- *Control:* Immediate diagnosis and treatment with antibiotics.

CHAPTER 2

PRINCIPLES OF CONDITIONING AND BEHAVIOR MODIFICATION

MOTIVATION

Animals respond to the environment to fulfill their basic biological objectives, such as maintaining life and reproducing themselves. Animals do not perform basic behaviors like eating and mating because they feel the desire to maintain life or reproduce—they do so because nature has arranged matters so that it "feels good" to engage in these behaviors. When we train animals, we exploit the animal's desire to "feel good" by requiring the animal to do as we wish before we allow it to engage in one of these basic motivating behaviors. Our best way of measuring the strength of a motivation is to see how much effort and trouble an animal will go to in order to get the chance to engage in a specific behavior, like eating or playing.

Needs and Drives

Behavioral scientists have long tried to form theories that adequately describe and explain motivation. Along the way, they have employed terms like "instinct," "need," and "drive" to express the idea that animals preferentially engage in certain kinds of behavior, and even exert enormous effort to get the chance to do so. These terms are no longer considered to be valid, scientifically speaking, and science has moved on to other ways of dealing with motivation. However, it is perfectly adequate to speak of "needs" or "drives" when describing behavior for the purposes of dog training. Needs range from those that are clearly physiological, like thirst, to those that are a puzzle to us because they do not seem to fulfill any immediate biological requirement, for instance, the drive to engage in play behavior. In any case, no matter what the source of the drive or need we use to motivate the animal, much of dog training involves arranging matters so that the dog's desires are gratified when it behaves in desirable ways. However, before expecting the animal to learn and work, the handler must ensure the dog's primary needs are adequately met. The dog must be healthy and happy and feel an emotional bond with its trainer. The dog's needs and drives include the following:

Primary Drives
The expression "primary drives" will be used to refer to the motivations for those behaviors that function to prevent physiological or physical injuries.

Oxygen
Breathing is perhaps the dog's most immediate need. Exercise or excitement creates an increased oxygen requirement, which causes panting. Note that heavy panting may hinder the dog's olfactory ability. In addition, keep in mind that a dog that is panting heavily may be overheated and/or physically exhausted, and is not in a physiological state that is conducive to learning. Therefore the trainer should avoid working on new lessons or problem behaviors when the dog is fatigued.

Water
The trainer must provide adequate quantities of water to prevent thirst from interfering with learning or task performance. Do not use water as a reward in dog training.

Food
The trainer must supply adequate quantities of food to prevent hunger from interfering with task performance. You may use food as a reward. The majority of dogs have sufficient appetite so that they will work strenuously for extra food rewards, particularly when these rewards are highly palatable "treat" foods. Food deprivation is not required. Intense

physical exertion, particularly in hot conditions, should be avoided when the dog has recently eaten.

Pain avoidance

A dog will avoid objects and actions that it has learned to associate with pain or discomfort, and this behavior is frequently exploited by dog trainers. The use of a physical correction, however, does not necessarily teach a dog the correct response to any specific cue. The trainer cannot assume the dog "knows what it did wrong." In addition, natural defensive responses to corrections may often interfere with the target behavior unless the dog already understands the desired response (e.g., pulling downward on the choke collar to try to make the dog lie down may simply result in teaching the dog to brace its forelegs and strain upward, unless the dog has previously learned how to lie down to earn a reward, and then learned to "turn off" collar pressure by lying down). The dog must know the correct response before the handler can use training that depends upon the dog's desire to avoid discomfort.

Secondary Drives

In addition to the primary needs like food and pain avoidance, the dog has other behaviors that can be exploited by providing the trainer with ways to reinforce and reward its behavior.

Dog trainers sometimes speak of the dog's desire to socialize as the "pack drive." The handler must keep in mind that one of the dog's strong drives is to enjoy a stable social relationship with one or more other beings, to "belong" to someone. A predictable and stable relationship in which the dog trusts (and has affection for) its handler is the basis of any effective system of training. This relationship does not form instantly—the handler must take the time and trouble to foster it. A period of socialization ("rapport-building") between dog and handler is required to establish this social relationship, and to render verbal and physical praise from the handler rewarding to the dog.

Dominant or "alpha" socialization

In most cases, a dominant dog will strive to achieve rank in a pack or social group. This behavior is a normal part of the character of many working dogs. To work effectively with a highly dominant dog, the handler must gain the initiative in the relationship. However, this is not done simply by "showing the dog who is boss." Attempts to physically punish a dominant dog into cooperative behavior normally only results in handler aggression and the dog and handler becoming suspicious of one another.

Subdominant or "beta" socialization

A subdominant dog is driven to behave in affiliative ways that will establish its belonging in the pack. These affiliative responses are called "submissive behavior." However, keep in mind that a dog's social rank or dominance with respect to its handler is not an index to its quality, even as a controlled aggression dog. Many strong patrol dogs are compliant and submissive with their handlers but are capable of very strong aggression toward "outsiders" when commanded. A submissive dog will often exhibit what is called "willingness" or "eagerness to please." This behavior can greatly facilitate the dog's training if the handler has established a positive rapport with the animal. Especially in the case of a submissive dog, excessive corrections and compulsion (see "Primary and Secondary Reinforcement and Punishment" section below), whether verbal or physical, can degrade rapport and decrease the dog team's proficiency in training and deployment.

Play socialization

Play is difficult to define precisely, and although scientists argue about what its purpose is, play is a distinct and identifiable behavior that occurs in a very wide range of animals. We may presume that it is one of the dog's needs, and there can be little doubt that carefree and happy play between dog and handler is a vital part of a healthy and productive training relationship.

- *Prey.* "Prey drive" is an expression that refers to the dog's natural tendency to chase, bite, and carry an item the dog perceives as prey. This applies to things that would, in the natural world, constitute prey items for a dog (e.g., a rabbit), as well as artificial objects (e.g., a thrown ball) that are also capable of triggering the dog's impulse to engage in predatory behavior. Prey behavior has enormous importance for the training of MWDs because it provides the reinforcement for nearly all substance detection training, and it also contributes very importantly to controlled aggression training. Many dogs display elements of social play behavior while retrieving balls and toys, and balls and toys can be thought of as play facilitators in addition to surrogate prey objects.

• *Aggression.* Even though there is not much evidence that animals have a "need" to behave aggressively, dog trainers still tend to speak of aggressive behavior as being based on one or more "drives." There are many different types of aggression, including dominant, defensive, and pain-elicited aggression. Aggression plays a vital role in MWD training and utilization because it is the foundation of patrol work. In addition, because MWDs are selected for a moderate to high level of aggressiveness, an MWD handler must at all times be aware of his dog's potential for aggressive behavior. The MWD handler must handle his dog responsibly and with care to prevent injuries to himself, to the dog, to other dogs, and to bystanders and co-workers. The handler must also help to prevent the development of handler-aggressiveness in his dog by: 1) treating the animal compassionately, humanely, and fairly; 2) avoiding a reliance on strict or overly compulsive methods for training (see "Primary and Secondary Reinforcement and Punishment" section below); 3) ensuring at all times that the dog clearly understands how to perform the desired skills; 4) renouncing emotionality—when you get angry and frustrated with your dog, put up and think the situation over!

LEARNING AND CONDITIONING

Learning is a permanent change in the behavior of an organism as a result of interaction with the environment. This definition distinguishes learning-based behavior change from other short-term behavior changes such as sensitization, fatigue, and sensory adaptation. The terms learning and conditioning are synonymous. For the purpose of dog training, it is sufficient to discuss three types of learning—habituation, classical conditioning, and instrumental conditioning.

Habituation
Habituation is a gradual decrease in the strength of responsiveness to a stimulus as a result of repeated experience with that stimulus. Think

of habituation as a mechanism by which organisms learn what to pay attention to and what to ignore. For instance, the first time a dog hears a door slam it may startle. In all likelihood, when it hears this slam again and again, it will gradually startle less and less until finally it exhibits very little response. This is adaptive because it allows the dog to save its energy and attention for more important events and stimuli, such as the noise of the lid coming off of the dog-food can. Habituation takes place continually during dog training, in ways that are both advantageous and disadvantageous to the dog trainer.

Advantageous habituation

Dogs are normally to some extent frightened of, or interested in, things that are new to them. However, MWDs are expected to carry out their duties in environments that feature very distracting and sometimes intense stimuli, such as taxiing airplanes and marching formations of personnel. Through habituation a working dog can learn to respond minimally to irrelevant stimuli and pay attention to its "job."

- *Procedures for advantageous habituation.* Habituation proceeds most effectively and rapidly with stimuli that are mild or moderate in intensity. (In fact, if a fearful dog is exposed repeatedly to a very intense stimulation, such as a running helicopter engine at close range, the animal is likely to respond *more* intensely to this stimulus over time instead of less intensely.) Habituation is most efficient when the stimulus exposures and training sessions are distributed or well-separated in time. For example, a dog can learn more easily to stop being startled by gunshots when the gunshots are spaced out at intervals of 15 or 20 seconds, and when the training sessions are separated by 24 hours. The opposite procedure, exposing the dog to rapid series of gunshots several times a day, may lead to increased fear of the noise. When habituation is conducted with mild stimuli on a distributed basis, undesirable responses such as fear tend to disappear more or less permanently.

- *Hierarchies of intensity.* Sometimes the trainer needs to cause a decrease in undesirable responding (i.e., fear) to a very intense stimulus, such as gunshots or jet engines, yet repeated exposure to these stimuli at full-strength is likely to produce even more undesirable responding. To cause habituation to these intense stimuli, it is necessary to modify them so that they become milder. A practical way to decrease the intensity of noise stimuli is by exposing the dog to them at a great distance. For instance, once the dog exhibits little fear to a jet engine at 500 yards, then the animal can be brought a little closer, and so on. The scale of noise intensity, ranging from very mild to very intense, is called a hierarchy. By introducing the dog to the stimulus at a level of intensity that is so low that it provokes very little or no fear responding, and by moving very slowly and gradually from one stage on the hierarchy of intensity to the next, the trainer may be able to teach his dog to eventually exhibit little fear in the presence of even very intense stimuli.

• *Counter conditioning.* Habituation processes can be made more powerful by exploiting another stimulus (e.g., food or ball) to offset the behavioral reaction (fear) caused by the stimulus we want the dog to be less sensitive to (e.g., jet engine). If we are far enough "down" the hierarchy—far enough away from the jet engine—the dog will become oblivious to the noise and intent on eating or playing. The pleasant emotional reactions to the food or ball will counter condition the jet engine, reducing the fear response to it. However, if we are too far "up" the hierarchy—too close to the frightening jet engine—we will instead find that the jet engine will counter condition the food or ball, reducing the dog's pleasurable response to these motivators. **Note:** Counter conditioning is actually a form of classical conditioning (see "Classical Conditioning" section below), but it is introduced here for the sake of clarity.

• *Spontaneous recovery.* Fear responses, especially, are very durable and persistent. They tend to reemerge even after extensive training. In fact, because habituation includes certain short-term processes that "wear off" after a few minutes or hours, it is normal for a habituated response to reappear to some extent between training sessions. Thus a dog may exhibit no fear of a stimulus by the end of one day's training session, yet show recovered fear at the beginning of the next day's session.

Disadvantageous habituation

In some circumstances, effective performance in a working dog also depends upon a certain level of interest in, and responsiveness to, environmental stimuli and routines. A dog that is relatively new to detection work, or obedience, or patrol training, may deliver very animated and lively performance because it is still stimulated and excited by these situations. The disinterested and inefficient performance that a dog handler often describes by saying "my dog is bored with the work" may be the result of habituation to training and deployment scenarios. This is disadvantageous habituation. To some extent, we can retard and offset disadvantageous habituation by changing the training scenarios constantly and offering the MWD as much variety in its daily work as possible (in addition to effective and timely positive reinforcement).

CLASSICAL CONDITIONING

In this form of learning (also called Pavlovian conditioning), the dog learns that there is a relationship between two events, or stimuli. One of these stimuli is a "neutral" or unimportant stimulus like the ringing of a bell—something that a dog would normally pay little attention to. This stimulus is called the *conditioned stimulus*, or CS, because it can generate strong behavior only as a result of conditioning. The other stimulus is a biologically important stimulus that a dog naturally pays a lot of attention to—like food. This stimulus is called the *unconditioned stimulus*, or US, because it can generate strong behavioral responses without any conditioning. Thus, a dog normally responds to the ringing of a small bell by merely perking its ears or looking toward the noise. However, a piece of food can cause the dog to show a great deal of strong behavior like excitement, salivation, digging and pawing, chewing, and eating. This very strong behavior caused by exposure to a US like food is called the *unconditioned response*, or UR. Through classical conditioning, the CS and the US become associated in the dog's "mind," so that the behavior that is naturally triggered by the US (the UR) comes to be triggered by the CS also. When a CS develops the ability to trigger behavior that is normally caused by a US, this learned response is called the *conditioned response*, or CR. In the classical example, the Russian scientist Ivan Pavlov taught dogs to salivate in response to the ringing of a bell. Pavlov did this by repeatedly pairing the bell (CS) and the food (US), presenting them close together in time. Eventually the dog learned that the bell predicted food, and then it began to salivate when it heard the bell (CR).

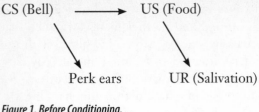

Figure 1. Before Conditioning.

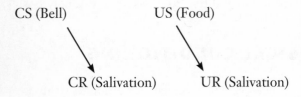

Figure 2. After Conditioning.

Classical Conditioning Procedures

Normally, the most effective way to "condition" a CS, to associate it with a biologically potent US, is to present the CS and then follow it very quickly (within a second or less) with the US. Thus, if the handler wishes to train his dog to feel startled and anxious in response to the word "No!" then an effective method would be to wait until the dog engages in some misbehavior like sniffing the trash. The handler would then give the "No!" cue, and throw a chain choke collar into the side of the trash can so that it makes an unpleasant sound about ½ second after the "No!" Originally the word "No!" (CS) will mean little to the dog and produce little change in behavior. The unpleasant noise (US) will be potent and cause a strong startle or freezing response (UR). Pairing the "No!" with the unpleasant noise will condition startling to "No!" (CR) within a very few CS–US pairings.

Then, when the dog is engaged in misbehavior, the handler can use the "No!" command, the dog will freeze or startle (which serves to interrupt the undesirable activity), and the handler can then call the dog to him and praise it. The dog will soon learn to shy away from behaviors and objects when it hears the "No!" command (classical conditioning) and return to its handler for praise (see "Instrumental Conditioning" below).

Backward Conditioning

When the CS and the US are reversed, so that the US actually occurs before the CS, this is called a backward conditioning procedure. Little or no learning takes place during backward conditioning. Thus, in the above example, if you first startled the dog with a loud noise and then said "No!" you would find that you could do this many times and still the dog would not show a startle response when "No!" was given by itself.

Importance of Classical Conditioning

Very little of working dog training involves the deliberate creation of classically conditioned associations, like the above example. Most of the "action" in dog training has to do with the use of reinforcers and punishers in instrumental conditioning (see "Instrumental Conditioning" below). However, it is still very important to understand classical conditioning processes because they underlie almost everything that takes place in dog training. Classically conditioned associations help the trainer and contribute positively to training in countless ways. For instance, if a handler makes an announcement ("This is SSGT Smith of the . . ..") prior to sending his Patrol MWD into a building to search for and find a hidden agitator, the dog will associate the sound of the announcement (CS) with the aggressive cues and behaviors that it experiences shortly thereafter (US) when it finds the agitator and bites. It will begin to exhibit aggressive responses to its handler's announcement—excitement and barking (CR)—that help to prepare it for the search and the bite. However, classically conditioned associations may also interfere with training. For instance, if the handler decides that his dog sits too slowly in response to the command, he may decide to hasten the sit by applying physical force. The handler gives the command "Sit!" in a loud voice, watches for a moment to see if the dog is sitting, and then gives a strong jerk upward on the choke chain. The handler intends to demonstrate to the dog the consequence of sitting slowly, but unwittingly he actually constructs a very effective classical conditioning trial "Sit!" is immediately followed by a sharp jerk on the collar. Soon "Sit!" develops the power to trigger responses that are normally only triggered by a sharp collar correction. These responses can range from pain-elicited aggression and biting to avoidance and cowering. More commonly, they involve anxiety and a reflexive stiffening of the body's muscles to defend against the sharp jerk on the collar. Thus, although the handler intends to speed the sit up, he is actually teaching the dog to respond to the "Sit!" command with anxiety and stiffening. The physical stiffening hinders the dog's ability to sit quickly, with the result that it receives yet another jerk on the neck, which makes it even more anxious and stiff when it hears "Sit!" and so on.

Extinction of Classically Conditioned Behavior

To make a classically conditioned behavior disappear, what we must do is present the CS repeatedly over and over again without pairing it with the US. The CR will gradually decrease in strength until it disappears. This procedure is called extinction. It is just like habituation procedures, except in habituation we are getting rid of an unlearned response, whereas in extinction we are getting rid of a learned response. Keep in mind that, just because a learned behavior has been extinguished, this does not mean that it has been unlearned or "erased." There is much evidence that learning causes permanent changes in the brain that are not reversed by extinction.

INSTRUMENTAL CONDITIONING

Instrumental conditioning and operant conditioning mean almost the same thing, except that instrumental conditioning is a slightly more general and flexible term. Instrumental conditioning refers to the way that rewards and punishments change the strength, or probability of occurrence, of prior behavior. Another way to put this is to say that behavior is modified by its consequences. Thus if a dog engages in a particular behavior such as investigating an odor, and then it encounters food, the odor-investigation behavior will be more likely to occur again in the future, and/or stronger when it does occur. This is an example of reinforcement. On the other hand, if a dog investigates an odor, and then it receives a jerk on the collar from its handler, the odor-investigation behavior will be less likely to occur in the future, and/or weaker when it does occur. This is an example of punishment.

Distinction Between Classical Conditioning and Instrumental Conditioning

For the purposes of dog training, it is adequate to think of classical and instrumental conditioning as separate processes that can be distinguished from one another in the following ways: Classical conditioning involves learning that there is a relationship between two environmental events, or stimuli, such as the ring of a bell and food, or the command "No!" and an unpleasant event. Instrumental conditioning mainly involves the animal learning that there is a relationship between its own behavior and some stimulus, such as the act of sitting and praise from the handler, or the act of searching for odor and a ball. Classical conditioning effects mainly what are called autonomic responses; things like reflexes and feelings and emotions that are not under the dog's voluntary control. Instrumental conditioning effects mainly what are called skeletal responses; behaviors like sitting, running, standing, and biting that are under the dog's voluntary control.

Stimulus/CS (Bell) _____Association_____ **Stimulus/US** (Food)
     CR—Salivation (involuntary behavior)

Figure 3. Classical conditioning.

Response (Sit) _____Association_____ **Stimulus/**Consequence (Food)
     Instrumental response—Sit (voluntary behavior)

Figure 4. Instrumental conditioning.

Response Contingency

Contingency is a term that refers to a relationship between two occurrences. When we say that one event is contingent on another, that means that one event will not occur unless the other occurs. In instrumental conditioning there is a contingency between a particular behavior, or response, and a stimulus. Thus a handler will not give his dog food unless the animal first sits. The relationship between sitting and food in this example is called a positive response contingency. This is a final and very important distinction between classical and instrumental conditioning—in classical conditioning there is no response contingency. Pavlov did not ring the bell and wait for the dog to salivate before rewarding the animal with food, he merely rang the bell and then presented food, regardless of what the dog did.

Positive response contingency

This is a relationship between a response/behavior and a specified event such that if the behavior occurs it will be followed by the event. For instance, if the dog sits, its handler will give it food. On the other hand, if the dog tries to bite another dog, its handler will give it a jerk on the leash. Even though this last example does not sound "positive" because it is not pleasant for the dog, it is still a positive response contingency. In the language of learning, "positive" is not used to refer to pleasantness. It is used to say that one thing will happen provided that another happens first.

Negative response contingency

A relationship between a behavior and specified event such that if the behavior occurs it will NOT be followed by the event. For instance, if the dog sits, its handler will not give it a jerk on the leash. Similarly, if the dog fails to find a hidden training aid, its handler will not allow it to have the ball. Although this last example does not sound "negative" because it does not involve anything unpleasant happening to the dog, it is still referred to as a negative response contingency. In the language of learning, "negative" is not used to refer to unpleasantness. It is used to say that one thing will not happen if another happens first.

Consequence

A consequence is an event that happens to the dog after it performs some instrumental behavior. There are two main categories of consequence (reinforcement and punishment). When we combine these two types of consequence with the two types of response contingency (positive and negative), we get four possible consequences that can result from any instrumental behavior—positive and negative reinforcement, and positive and negative punishment (see "Inducive Training" below).

Reinforcement

A reinforcer is an event that encourages or strengthens prior behavior. Examples of reinforcers are food, access to a toy, or a pat on the head. Any of these, when given to the dog after it sits, tends to strengthen sitting behavior. Food, toys, and pats on the head are reinforcing because they are pleasant. However, unpleasant events can also act as reinforcers.

The handler can reinforce a behavior with an unpleasant event like a jerk on the leash by withholding the jerk when the dog sits. In this example, there is a negative response contingency between sitting behavior and a jerk on the collar—if the dog sits, there will be no jerk. Although the jerk itself is unpleasant, the absence of the jerk is a "satisfying state of affairs" and will, under proper circumstances, serve to reinforce sitting behavior.

- *Positive reinforcement or reward.* Positive reinforcement is the use of intrin- sically pleasant stimuli like food, toys, and pats on the head to strengthen and encourage prior behavior. The word "positive" does not refer to the "goodness" or "pleasantness" of the stimuli, it instead refers to the positive response contingency between a target behavior (like sitting) and the reinforcing event—if the dog sits, it will be given food, toy, or a pat on the head. Positive reinforcement is synonymous with reward.

- *Negative reinforcement.* Negative reinforcement is the strengthening of behavior by using the withholding or withdrawal of intrinsically unpleas- ant events like jerks on the collar. The word "negative" does not refer to the "badness" or "unpleasantness" of these stimuli, instead it refers to the negative response contingency between a target behavior (like sitting) and some event—if the dog sits, it will NOT be given a jerk on the collar.

Punishment

A punishment is an event that discourages or weakens prior behavior. Examples of punishers are jerks on the collar (collar corrections) or a bump on the nose. Either of these, when administered to a dog after it misbehaves by, for example, departing from the down-stay position without permission, will tend to weaken down-stay–breaking behav- ior. Collar corrections and bumps on the nose are punishing because they are unpleasant. However, pleasant events can also act as punish- ers. The handler can punish an undesirable behavior by withholding or taking away a pleasant stimulus like praise and petting. Thus, if the dog tends to jump up on his handler when it is excited, seeking attention, this behavior can be punished by withholding praise and attention. In this example, there is a negative response contingency between jumping-up behavior and praise and petting—if the dog jumps up, it will not receive praise or petting. Although the praise and petting are themselves pleasant, their absence is an "unsatisfying state of affairs" and will, under the proper circumstances, serve to punish jumping-up behavior.

- *Positive punishment or punishment.* Positive punishment is the use of intrinsically unpleasant stimuli like collar corrections to discourage or weaken behavior. The word "positive" does not refer to the "pleasantness" or "unpleasantness" of the stimuli, it instead refers to the positive response contingency between a target behavior (like breaking the down-stay) and the punishing event—if the dog breaks the stay, it will be given a collar correction. To simplify, we can use the simpler term "punishment" in place of "positive punishment."

- *Negative punishment, or omission.* Negative punishment is the weakening or discouragement of prior behavior by withholding pleasant events like food or praise and petting. The word "negative" does not refer to the "pleasantness" or "unpleasantness" of these stimuli, instead it refers to the nature of the negative response contingency between a target behavior (like jumping up) and an event—if the dog jumps up, it will NOT be given praise and petting. To simplify, we can use the expression "omission" to refer to "negative punishment."

Contingency Square

The contingency square is a table that graphically depicts the relationships between reinforcement and punishment (i.e., the effect the instrumental procedure has on behavior) and the nature of the response contingency (positive and negative—the handler gives something to the dog or the handler withholds something from the dog). Memorizing the table will help the trainer remember each of the four possible consequences of an instrumental behavior, and their definitions. For example, take negative reinforcement, normally the most difficult of the consequences for trainers to understand. The key is to take each of the words of the term negative reinforcement and analyze it separately to understand whether the consequence involves pleasant or unpleasant events for the dog. "Negative" means a negative response contingency—the handler will withhold something or take something away if the dog performs a target behavior. "Reinforcement" means that the outcome will be to encourage or strengthen the target behavior. What must I withhold or withdraw from a dog to encourage prior behavior? An unpleasant event.

Therefore, to use negative reinforcement means to encourage a dog's behavior by removing or withholding from the animal something that it does not like. For instance, we can reinforce a dog for dropping a ball by releasing pressure exerted on his neck with a choke chain. Reward and omission DO NOT involve the use of force or discomfort to modify the dog's behavior. Negative reinforcement and punishment DO involve the use of force or discomfort to modify the dog's behavior.

Table 5. Nature of the Response Contingency

INDUCIVE	COMPULSION
POSITIVE (WITH) REINFORCEMENT Use of pleasant stimuli, such as food, toys, and petting	POSITIVE (WITH) PUNISHMENT Use of unpleasant stimuli, such as a collar correction
NEGATIVE (WITHOUT) PUNISHMENT Withholding or absence of pleasant events, such as food or praise	NEGATIVE (WITHOUT) REINFORCEMENT Reinforcement of behavior by withholding compulsion

Compulsive Training

Compulsion is a word that refers to forcing or coercing people or animals to do things. In compulsive dog training, the handler relies on unpleasant events to obtain desired behavior from the dog. Thus, compulsive training involves the use of negative reinforcement (encouraging desirable behavior by withdrawing or withholding unpleasant stimuli) and punishment (discouraging undesirable behavior by administering unpleasant stimuli). Although the training of working dogs often involves the use of some compulsive methods, it is important to understand that: 1) These methods are effective and humane only under certain circumstances—when the dog is well-prepared and already understands the desired response and how to avoid compulsion; 2) Excessive reliance on compulsion will damage the dog's rapport with its handler and cause it to dislike and avoid work; 3) Compulsion may stimulate defensive and aggressive responses in the dog, and can in many circumstances be counterproductive and even dangerous for the handler; and 4) Some phases of working dog training, most especially the detection phase, are incompatible with compulsive techniques.

Inducive Training

Inducive training is the opposite of compulsive training. The root word "induce" means to gently persuade. In inducive training the handler relies on the use of pleasant events and stimuli to obtain desirable behavior from the dog. Thus, inducive training involves the use of reward (encouraging desirable behavior by administering positive reinforcement) and omission (discouraging undesirable behavior by withholding or withdrawing positive reinforcement).

Primary and Secondary Reinforcement and Punishment

Many rewards and punishments are stimuli that are biologically powerful, such as food or pain. In the language of classical conditioning, they are called unconditioned stimuli (USs). In the language of instrumental conditioning, they are called primary reinforcers or primary punishers. Dogs respond readily and strongly to these stimuli without having to be taught to do so. However, some rewards and punishments are things, such as the words "Good!" and "No!" that originally have little effect on a dog's behavior. These are called secondary reinforcers and punishers because they do not become effective until they have been associated with primary reinforcers or punishers.

Secondary reinforcers

Secondary reinforcers gain their pleasant value by being associated with primary reinforcers. For instance, puppies probably do not instinctively enjoy being spoken to. They learn to like being spoken to in a happy voice because this voice is associated (through classical conditioning) with physical petting and with the presentation of food. After enough of this conditioning, words like "Good!" spoken in a happy voice become pleasant stimuli. Subsequently, the word "Good!" has the power to reinforce prior behavior (if the handler says "Good!" immediately after the dog executes the behavior).

Secondary punishers

Secondary punishers gain their unpleasant value by being associated with primary punishers. For instance, the word "No!" means nothing to an untrained dog. The word becomes unpleasant because it is associated (through classical conditioning) with unpleasant primary punishing events like a jerk on the collar. After enough of this conditioning, the command "No!" spoken in a stern voice becomes an unpleasant stimulus. Subsequently, the word "No!" has the power to punish prior behavior (if the handler says "No!" immediately after the dog executes the behavior).

DISCRIMINATIVE STIMULI

Thus far in our discussion of instrumental conditioning we have described only the contingent relationship between a target behavior and a reinforcer or a punisher (e.g., sit–food, or jump up–"No!"). However, to behave appropriately in training, the dog must know when these contingent relations are actually in force. The handler will not reward a sit anytime the dog sits, but only when he desires the dog to sit. The way the handler signals to the dog that he wants it to sit is with the command "Sit!" This command tells the dog that now one or more response contingencies are in force—for instance, a prompt sit will result in petting and praise and the omission of a collar correction, while refusing to sit will result in no petting or praise and the administration of a collar correction. Thus, our full model for the use of instrumental conditioning can be symbolized as follows: $Stimulus^{Discriminative}$–Response–Consequence, or S^D-R-C. S^D is the command "Sit!," while R is the dog's response (sitting or refusing to do so), and C is the consequence of the dog's behavior (reward, negative reinforcement, punishment, or omission—food, petting and praise, collar corrections, etc.). This three-term model shows that the dog must actually learn at least two associations for any command skill—one between the behavior and the consequence, and one between the command and the behavior. In some types of animal training, these two associations are taught separately. For instance, first a killer whale is taught to jump for reinforcement, and then it is taught that the jump-reinforcement contingency is in force only after the trainer issues a command. If the whale jumps at any other time, no reinforcement will be forthcoming. However, in dog training both associations are normally taught simultaneously because the handler always includes the command in lessons (see "Application of Inducive Training" below).

INDUCIVE VERSUS COMPULSIVE TRAINING

Some compulsion is normally necessary in working dog training, especially in the controlled aggression phase. However, inducive methods are to be preferred whenever practical. In particular, inducive methods are most advantageous for the initial teaching of any skill. That is, to an untrained dog the SD/command (e.g., "Sit!") means nothing. Therefore, if the handler gives the command "Sit!" and then administers a strong collar correction in the attempt to force the dog to sit, the dog will have no idea that it can avoid further unpleasantness by sitting. It will instead attempt to defend itself or avoid its handler. (More than anything else, such a method is a perfectly designed classical conditioning procedure that will condition fear and/or aggression to the command "Sit!" by pairing the command closely together in time with physical discomfort.) However, if we first teach the dog to sit on command using inducive methods, and ensure that it understands what sit means and that it has learned to enjoy training, then we may constructively use compulsion to hasten the dog's sit or to teach it to sit even in distracting circumstances. Thus, the proper role of inducive training is to teach the dog skills, while the proper role of compulsive training is to enforce the performance of these skills (if necessary). In addition, another very important role for inducive techniques is during the early stages of handling any dog, even a well-trained dog. The optimal way for a handler to build rapport and a good working relationship with a new dog is to perform inducive training exercises with the animal (e.g., food–rewarded sits), even if the dog already knows these exercises, and avoid the use of physical compulsion.

APPLICATION OF INDUCIVE TRAINING

In inducive training, the handler employs gentle means to lead a dog to perform some target behavior, and then he reinforces this behavior. In the event that the dog does not execute the desired behavior, or executes it incorrectly, then the handler will omit reinforcement (see "Negative Punishment or Omission" under "Punishment" above). In the classical example, the handler teaches a dog to sit by drawing the dog's attention to a piece of food in his hand. Once the dog places its muzzle in contact with the handler's fist in the attempt to take the food, the handler then slowly raises his hand and moves it slightly backward toward the dog's tail, simultaneously giving the command "Sit!" In following the movement with its head, the dog is very likely to sit. The handler then opens his hand to feed the dog and administers praise. If the dog fails to sit, the handler withholds reinforcement and praise (omission) and continues attempting to "finesse" the sit. In such a stress-free setting, a dog can learn very rapidly to sit on command. In addition, it also learns to enjoy work, and it develops affection and trust for its handler. There are similar methods for teaching almost all of the exercises of obedience.

Successive Approximation and Shaping

Successive approximation is a practice in which animals are taught behaviors by rewarding responses that are progressively more and more like the desired target response. For instance, to teach a dog to sit through successive approximation, a handler would wait until he observed a tiny approximation of a sit on the dog's part, such as flexing of the hind legs, and then reinforce this movement. Once the dog flexes its legs readily for reinforcement, then the handler would withhold reinforcement until the animal exhibited a flexing that was slightly greater than before, that approximated a little more closely an actual sit, and then he would reward this behavior, and so on. The entire process of extracting a trained response through successive approximation is called behavior shaping. Successive approximation and shaping are of vital importance in the training of exotic animals such as killer whales and sea lions, but they play comparatively little role in dog training, for the simple reason that a good dog trainer can usually think of a way to get the dog to offer the complete behavior, and then reward that, as described in relation to the sit in "Application of Inducive Training" above. However, particularly in the case of very complex or difficult behaviors, or behaviors for which the dog is handicapped or contra-prepared (i.e., when a dog has a history of problems with a particular exercise), it is very important to realize that often a good handler will reinforce his dog for a good "effort" in the direction of the desired target behavior. This will encourage the animal and lead it to continue trying to learn the lesson.

Reward Schedules

A reward schedule is a rule that dictates how often a dog will receive positive reinforcement when it correctly executes a skill. It is very important to understand these schedules, because they produce different effects and are appropriate at different stages of the training of each skill. There are six types of reward schedule we should consider:

Extinction schedule

To extinguish an instrumental response, we merely allow the behavior to occur again and again, without rewarding it. The behavior will gradually decrease in strength and frequency until it disappears. Thus, to extinguish an undesirable behavior like jumping-up, it is often sufficient to identify the reward for the behavior (it is usually some reaction given by the handler when the dog jumps up) and then make sure that this reward never follows the problem behavior. This is called "putting jumping up on an extinction schedule." It is important to realize that some behaviors are "intrinsically" reinforcing—that is, just doing them is rewarding to the dog. If a behavior is intrinsically reinforcing, it will not extinguish even though we put it on an extinction schedule. Thus, if an anxious dog finds a way to release tension by barking in its kennel, it may not ever stop barking in the kennel, even if its handler is careful to never go to it when it is barking.

Continuous reward schedule (CRS)

Positive reinforcement is given immediately when the dog makes a correct (or sometimes a near-correct) response. Assisting the dog to assume the desired position or behavior is permissible (i.e., in the case of the sit, gentle pressure on the rump to encourage the animal to sit), but it is preferable to "finesse" the dog into the sit by baiting it with food or some similar technique. Inducing the animal to perform the desired behavior independently and then reinforcing the behavior will produce more rapid learning than "pushing" the animal into position and then rewarding it for allowing this to happen. CRS is the most effective reinforcement schedule for teaching a dog a skill.

Fixed ratio reward schedule (FRRS)

Positive reinforcement is given to the dog after it makes two or more correct responses. It is most useful to think in terms of ratio schedules of reinforcement in the case of behaviors that are "episodic" like barks and scratches. To start a dog on the FRRS schedule, every second response is rewarded. When the dog consistently makes two responses to obtain a reward, require three responses. By increasing the number of responses, one at a time, and allowing the dog to perform at each level with 100% proficiency, you can work up to a high FRRS. If the proficiency falls below 100%, then decrease the number of responses required to obtain a reward until the dog recovers its proficiency. Then proceed as before, adding one response at a time. A fixed ratio reward schedule is the way that a handler can best train his dog to, for instance, bark or scratch repeatedly at a door to indicate the presence of an agitator in a building. Initially, while we are trying to teach the behavior, the handler will open the door and allow the dog to bite (reward) after one bark or scratch, then he will require two barks or scratches before rewarding the dog, then three, and so forth.

Variable ratio reward schedule (VRRS)
Once the dog has learned to perform the maximum number of responses by the FRRS schedule, then use the VRRS. Select a range of responses (e.g., 5 to 10 correct responses) required and reward the dog on a random basis within this range (e.g., the dog has already learned to bark 15 times to obtain a bite on an FRRS). Now you should begin rewarding the animal somewhere between 5 and 10 barks—on a random basis, so that the dog never knows whether it will have to bark 5, 6, 7, 8, 9, or 10 times to get a bite. The dog will learn that it must correctly respond at least 5 times, and perhaps up to 10 times to obtain a desired reward.

Fixed interval reward schedule (FIRS)
Reinforcement is given to the dog after it responds to a command for a given fixed period of time. It is most useful to think in terms of interval schedules of reinforcement in the case of behaviors that are "continuous," such as staying in position, heeling, and searching. In initial training, select a short period. If the dog does not respond correctly, select a shorter period of time until the dog responds correctly to obtain a reward. As in the FRRS, add short periods of time (e.g., 5 seconds) to the interval and require the dog to attain 100% proficiency at each interval.

If the dog fails to respond correctly for the required period of time, readjust the time requirement to a lower time requirement until the animal regains 100% accuracy, and then begin gradually again to increase the required interval. Excellent examples are staying in a position (like the down) and walking at heel. In each of these cases, a good trainer initially rewards the dog for just a few moments of good responding. With time and practice the handler gradually extends the period of time that the dog must remain in the down, or walk at its handler's side. Gradual extension of the period of time a dog works before reinforcement plays a vitally important role in detection training. A good trainer arranges a search problem for a novice dog so that it can easily find the target odor and obtain reward in perhaps less than a minute, whereas he requires an advanced dog to work for 5 or 10 minutes or more prior to finding the target odor.

Variable interval reward schedule (VIRS)

Once the dog has learned to perform a task for a period of time on a FIRS, use the VIRS. Select a time range (e.g., 1 to 2 minutes) and reward the dog on a random basis within this time period. For example, if the dog has already learned to hold a down-stay for 3 minutes on a FIRS, then begin rewarding it somewhere between 1 and 2 minutes on a random basis. The dog will learn that it must hold the down for at least 1 minute and perhaps for up to 2 minutes to obtain reward.

Application of reward schedules

Normally, in dog training it is not necessary to exactly follow the above steps to get good results. It is usually sufficient to follow these general rules: When teaching a dog to give an episodic response (e.g., bark) begin by rewarding it every time it barks (CRS), then reward it gradually for longer and longer sequences of barking, working your way up to the maximum number of barks that will be useful (FRRS). At any point that the dog shows hesitation or confusion, decrease the number of barks required so that the animal regains proficiency and then begin working back up again. Once the dog barks rapidly and confidently about the maximum number of times desired to obtain its reward, then begin giving it rewards randomly for some number of barks less than the maximum (VRRS). When teaching a dog to perform a continuous response (e.g., holding a down-stay), establish performance by rewarding it consistently after a very short period of time (FIRS). Then begin gradually extending the period of time that you require the dog to stay. Do not hesitate to decrease the duration requirement if the dog's performance deteriorates. Once the dog stays solidly for about the maximum desired period of time, begin rewarding it randomly for stays of various durations less than the maximum (VIRS).

Advantage of variable reward schedules

You may ask "Why bother to use VRRS and VIRS schedules?" Using FRRS and FIRS, the animal has already learned to bark many times in succession or stay for several minutes. The reason is that variable schedules teach the dog to be persistent and stubborn in trying to obtain its reward through instrumental behavior. Many scientific studies have shown that variable reward schedules produce stronger and more persistent conditioned behavior than fixed schedules. The psychological basis of this persistence effect is well understood, but it is very complicated. So a good simple way to think of the variable reinforcement phenomenon is this: When the dog never knows how many times or how long it will be required to perform before being rewarded, it "loses track" of how many and how long and just concentrates on performing persistently for reward, convinced that if it tries hard enough it will eventually get what it wants.

APPLICATION OF COMPULSIVE TRAINING

Just as it is important to understand certain basic principles (such as reward schedules) to perform effective inducive training, it is also important to understand certain basic principles to use compulsive training effectively.

Use of Positive Punishment

Positive punishment is used to teach a dog not to do something. Of course, this doesn't mean that the dog should do nothing, but that it should do something else, such as sit still. There are four major principles the trainer must understand to use punishment effectively and humanely: 1) The dog must have the ability to perform the alternative behavior. For instance, if a dog is breaking the down-stay because it is frightened of a jet engine, the dog's fear may render it unable to do what is necessary to avoid punishment. That is, if a trainer physically punishes a frightened dog for not staying, the punishment is likely to make the dog even more afraid and less capable of staying. This is not fair nor humane nor effective dog training. 2) Do not "ramp up" corrections. That is, do not begin punishment by using a very soft correction, and then gradually increase it as needed. Dogs, especially very excited dogs intent on working their way to a reward, adapt quickly to physical punishment and can learn in a short period of time to endure very uncomfortable events without altering their behavior. It is possible to, without meaning to, create a "monster," a highly excited and stressed animal that can absorb enormous amounts of physical discomfort without changing its behavior into the desired path. Instead, begin punishment training with a correction of an intensity that is meaningful to that dog and sufficient to cause it to change its behavior immediately. 3) Do not use punishment if it is not working. That is, if you have tried to intervene with a problem behavior by using what you believe is a meaningful intensity of punishment for that dog, and the desired result is not achieved, think carefully before you apply stronger physical punishment. The dog may be, for any number of reasons, incapable of the alternative behavior. The dog may have a history of bad training that has rendered humane and reasonable levels of physical punishment ineffective. You may be making some errors in technique that are preventing a humane and reasonable level of punishment from having the desired effect. In any of these cases, it is inexcusable to continue to physically punish a dog. 4) Avoid emotion when administering punishment. If you are angry, or frustrated, or upset while administering punishment to a dog, you can be almost certain that you are making mistakes and being unfair to the dog. Revenge and temper tantrums have absolutely no place in working dog training—you must not let training turn into a spectacle of one dumb animal hurting another.

Use of Negative Reinforcement
Negative reinforcement is the reinforcing of behavior by withholding compulsion. The classic example in MWD training is the "Out!" in which the dog releases an agitator on command. Although a clever

handler uses whatever positive reinforcement he can to reward the dog for releasing cleanly (e.g., praise, immediate rebite, etc.), the "Out" is normally taught and maintained principally through the administration of negative reinforcement. Thus, if the dog releases cleanly on command, it will NOT be corrected with a jerk or pull on the choke collar. All of the principles stated above that apply to positive punishment apply to negative reinforcement as well. In addition, it is also vital to understand the following terms and definitions:

Escape training

Escape is an initial stage of negative reinforcement training. During this stage, the command "Out!" is meaningless. The dog does not yet understand that the command "Out!" means that if it does not release it will receive a collar correction. On the first trial, when the handler gives the "Out!" command and the dog continues biting, the handler then applies a collar correction until the dog releases the bite, praising the dog once it has released. In all likelihood one or several more trials will proceed much the same. Although the dog may not be releasing on command, it is learning all the same. During this stage the dog learns to expect the correction when it hears the command "Out!," and it also learns to "turn off" or terminate the correction once it is applied by releasing the bite. This escape learning is very important. A dog that does not know precisely how it can "turn off" compulsion will be stressed and upset by corrections, and may engage in inappropriate behaviors to try and terminate discomfort, such as biting its handler. This point is especially important when the escape behavior, the behavior that we desire to teach the dog, involves a complex response like walking at heel or recalling to heel. If these exercises are taught using negative reinforcement, there must necessarily be a stage during which the handler teaches the dog to terminate collar corrections by placing itself at heel. If the animal does not know how to terminate compulsion by placing itself at heel, then collar corrections will only make it move more and more strongly away from its handler. In fact, this is why it is so important to patiently teach the dog as many skills as possible by means of positive reinforcement prior to polishing any of them with negative reinforcement—to make sure that the dog knows how to perform all behaviors on command. In this way the dog is well prepared to learn very quickly, with minimal stress or confusion, how to terminate compulsion by executing a commanded behavior.

Avoidance training

Avoidance is the next stage of negative reinforcement training, during which the dog learns that, in addition to terminating compulsion by releasing the bite, it can also completely avoid compulsion. That is, if the dog releases the bite quickly on command, the collar correction will never occur. When avoidance is completely and cleanly taught, every time the dog releases on command, it is reinforced by the absence of the correction, as though the dog "beat the rap." Note that sometimes the dog, especially if it has prior experience with the "Out!" command, will go directly to avoidance (releasing on command) after only one correction, without a noticeable escape stage of learning.

Criterion avoidance

The end goal of negative reinforcement training is to secure correct response to the command every time, without the need to use compulsion to "escape" the dog into the desired behavior. In working dog training, this goal has the additional dimension that the handler also is training toward the point at which he can discard the means of compulsion (i.e., collar and leash). That is, a dog that is fully-trained to "out" not only releases cleanly on command, it also releases when the collar is not attached to the leash, and when the handler is 20 or 30 yards away. In these cases, the handler has given up his option to correct the dog effectively. If the animal fails to obey the command, the handler has no good options. This means that the handler must not discard the means of compulsion until the dog has achieved a good avoidance criterion—clean avoidance of compulsion by good response to command consistently and repeatedly over at least 4 or 5 training sessions. During these error-free training sessions, the handler stands ready to correct the dog instantly, with all necessary things in place, but does not ever need to. On the other hand, if the dog still occasionally fails to respond correctly, "testing" to see if the handler is ready to apply the correction, then the dog has not yet achieved criterion avoidance and it is not yet ready to proceed further in training. The animal must continue to practice, with the handler standing by, ready to enforce obedience, until criterion avoidance is obtained.

Supporting negative reinforcement with positive reinforcement

Although behavior learned through negative reinforcement training can be very durable and reliable, it is advisable to, whenever possible, support negative reinforcement with positive reinforcement—give the dog rewards in addition to the reinforcement of not being corrected. For instance, after a clean, fast out from the agitator, you might praise your dog quickly and then immediately let it rebite and take the sleeve away from the agitator. After the last out of the training session, after the agitator runs away, you can reward your dog for its good compliance by letting it bite and carry a section of rubber hose or some other toy.

GENERALIZATION OF CLASSICAL AND INSTRUMENTAL CONDITIONING

Generalization is a process in which behavior that is learned in response to one stimulus is expressed to some degree in response to another stimulus. Generalization takes place with both classically conditioned and instrumentally conditioned behaviors, and the more similarity there is between two stimuli, the more generalization there will be from one to the other. Thus, a dog that has learned a strong startle response to the "No!" command may also startle and return to its handler when he says "Yo!" loudly to a friend. A dog that has learned to sit in response to one explosive odor, such as ammonia dynamite, may also sit in response to a similar nonexplosive odor, such as ammonia-based house-cleaning liquids. These are both examples of undesirable generalization, but generalization may also work in our favor. For instance, if you are incapacitated during a patrol deployment, but your well-trained dog also releases the bite in response to your partner's "out" command that is desirable generalization.

Context Generalization

Trained behaviors are not just controlled by CSs (classical) and SDs (instrumental). To some extent, they are also controlled by context.

Context is the word psychologists use to label all of the stimuli present in the conditioning situation other than the CSs and USs, the SDs and consequences. Context means the "environment." Context definitely participates in learning, and generalization from one context to another is rarely perfect. As a result, a dog that has learned to search and detect in a warehouse may also do so when it is taken to an office building, but its search and/or detection behavior is liable to be substantially different in the office building. To a degree, much of dog training consists of teaching the dog skills, and then trying to make these trained behaviors as independent as possible of the context, so that the dog will perform correctly anytime and anywhere. The best way to make trained behavior independent of the context is to train in as many different places and situations as possible (after the initial learning phase).

LEARNING TRANSFER

Transfer of learning is what takes place when the learning of one skill or command affects the learning of another skill or command. Transfer can be positive (favorable) or negative (unfavorable).

Positive Transfer
In positive transfer of learning, the fact that the dog has already learned to do one thing actually helps it to learn to do another. Thus, learning to sit in response to the "Sit!" command during obedience training transfers positively to detection training, helping the dog to learn to sit in response to odor. In fact, one of the main ingredients to good dog training is teaching each skill at such a time and in such a way that it helps the dog learn the next skill.

Negative Transfer
In negative transfer of learning, the fact that the dog has already learned to do one thing hinders it when it is trying to learn another. For instance, if your dog has already learned to scratch at a door to get through it and reach an agitator, this may transfer negatively to explosives detection training, making it more likely to "aggress" a training aid rather than sit cleanly.

ANTICIPATION

As a result of classical and instrumental conditioning, the dog learns to predict what will happen next during training. This knowledge of what is about to happen is accompanied by psychological and physiological changes that prepare the dog for upcoming action. Much the same thing happens to you when, riding in a car, you see the car ahead lock up its brakes. As a result of your anticipation of a collision, you brace yourself. When the dentist starts the motor in his drill, you will tend to wince and stiffen your body in anticipation of pain, even before you can feel the drill. The dog's anticipation of the events in dog training and its preparatory responses can help it learn. For instance, if your dog is having difficulty responding without assistance to odor during detection training, it sometimes helps to allow the animal to find a particular training aid two or three times running. Because the dog anticipates finding the same aid in the same place and sitting, it will be likely to respond quickly and completely without an assist, giving you the chance to reinforce this independent behavior. On the other hand, there are many circumstances in which anticipation can interfere with learning desired behavior. For instance, during the obedience exercises at the end of the leash (EOL), the dog is normally commanded to change position a few times (e.g., sit-down-sit) **without moving toward the handler,** and then recalled to the heel position and rewarded. However, if we practice this complete sequence of exercises many times, we may find the dog's knowledge of the routine interfering with the changes of position—the animal's anticipation of returning to the handler and being rewarded will cause it to creep forward instead of staying in place while moving from sit to down and back to sit. Controlled aggression is a situation where anticipations can be particularly crippling to progress. When in the intensely motivated state that pertains during controlled aggression, the dog's anticipations have tremendous power and can create intense interference with ongoing exercises. So, for instance, if you have brought your dog to the line of departure for the biting exercises, it will tend to become very excited and tense and focus its attention completely on the agitator, ready to respond explosively to the first cue sending it to bite. Its powerful anticipation of the bite and its preparatory responses will make it very difficult to, for instance, ask it to pay attention to you and walk at heel away from the agitator. It is inappropriate and ineffective in many of these circumstances to use compulsion to overcome the dog's anticipation and rigidity—physical discomfort is likely to make the dog even more tense and rigid and aggressive. What is necessary is to find ways to offset the dog's anticipations so that they do not interfere so strongly with training—for instance begin teaching the animal that the command to bite will often come when it looks at its handler or walks at heel with its handler away from the agitator. When this anticipation is formed, the dog will naturally begin to "ask" for the "Git 'em!" command by looking at and moving toward its handler.

Compartmentalization

The single most useful technique for dealing with anticipation and interference is to separate, or compartmentalize, exercises that interfere with each other. Thus, in the EOL example above, the thoughtful trainer will seldom recall his dog from EOL. Instead the trainer will place the animal EOL, run it through a few changes of position, pause, and then go to the dog and release it and reward it. In this way the dog does not anticipate a recall at the end of the EOL exercises, and therefore does not creep forward. To practice the recall from EOL, on a separate occasion the trainer will place the dog EOL and make it stay for a while, and then recall it, in this way keeping the changes of position and the recall compartmentalized and preventing interference.

CHAPTER 3

PATROL DOG TRAINING

OBEDIENCE COMMANDS

Give voice commands sharply, crisply, and in unison with the corresponding hand command. After the handler/dog team becomes proficient, you may give the commands and/or gestures independently. The commands start with the instructor, directed at the handler (e.g., Instructor, "SIT DOG, COMMAND"; Handler, "SIT," with a hand gesture).

Obedience Commands Beside the Dog
Teach all basic obedience commands first on leash with the dog at the handler's left side. These commands and correct responses start and end with the dog in the HEEL/SIT position.

HEEL
The initial command and response is "HEEL." There are two HEEL positions for the dog, one is for marching and the other is for the stationary HEEL/SIT. Whether marching or in the HEEL/SIT position, ensure the dog's right shoulder is even with the handler's left leg, and the dog's body is parallel to the handler's body. The dog should not forge ahead or lag behind.

- Give the verbal and manual "HEEL" when the handler starts forward movements, changes direction, and at one pace before coming to a halt. Give the hand gesture by slapping the left leg with the open left hand, while commanding "HEEL." When called to attention, give the command HEEL as the left foot strikes the ground. At the command "Forward MARCH," give the command "HEEL" with the first step forward. If a dog lags behind, coax the dog into the HEEL position (not jerked) by patting the left leg, snapping the fingers, calling the dog's name, or verbally encouraging the dog. On movements to the left, give the command "HEEL" after the handler's right foot begins to pivot. This prevents the dog from blocking the pivot movement. On movements to the right and the rear, give the command "HEEL" as the handler pivots. The dog can then assume the HEEL position before the movement is completed.

- *HEEL/SIT.* After the dog learns to walk in the HEEL position, it must learn to HEEL and then SIT in the HEEL position. Once the dog has learned the separate responses of HEEL and SIT, the next step is to teach the dog to SIT automatically in the HEEL position when stopped without further command.

SIT

When the instructor gives the command "SIT DOG, COMMAND," the handler gives the command "SIT," while grasping the leash several inches above the choke chain with the right hand. Place the palm of the left hand over the dog's hips with the fingers positioned at the base of the dog's tail, apply upward pressure on the leash while pushing down on the dog's hindquarters. As the training progresses, the dog should no longer require physical assistance.

In learning the command SIT, the dog may get slightly out of position. If this occurs gently reposition the dog. Every time the dog assumes the correct position praise the dog. Take care not to make praise excessive, because this may cause the dog to break position.

DOWN

When the instructor gives the command "DOWN DOG, COMMAND," and when the handler gives the command "DOWN," the dog must promptly lie parallel to the handler with its right shoulder in line with the handler's left foot.

- The handler introduces the command "DOWN" when the dog is in the HEEL/SIT position. Give the hand gesture along with the verbal command.

Some dogs may resist going down because it places them in an unnatural position. Therefore, use caution because the dog could bite the handler. The handler first bends down and grasps the leash just behind the snap or the choke chain ahead of the snap depending on how much space is needed to apply downward pressure on the leash. Then, while giving the DOWN command, apply pressure firmly toward the ground until the dog lies down. If the dog assumes the DOWN position without resistance, the handler should praise verbally before returning to the position of attention. Take care not to make praise excessive; this may cause the dog to break position. Use the command "STAY" before returning to the position of attention.

- To place a resisting dog in the DOWN position, the handler kneels down and grasps the leash just behind the snap with the left hand; then place the right arm behind the right front leg and grasp the left front leg about 6 inches above the foot. While pressing down on the leash, command "DOWN" and push the front legs forward until the dog is in the DOWN position.

- Once the dog has learned the DOWN command, you may need to correct the dog's position. If this occurs, give the command "SIT"; and after the dog sits, repeat the down process. Take care not to move the left foot while correcting the position because the dog is trained to line up on the left foot/leg.

STAY
The stay command is introduced while the dog is in the HEEL/SIT position and used for any position you commanded the dog to assume. Ensure the hand gesture is distinct, decisive, and executed in the following manner:

1. Lock the left arm at the elbow.
2. Turn the hand until the palm faces rear and open it until the fingers are extended and together.
3. Move the extended, locked arm forward until the arm and body make an angle of approximately 45 degrees.
4. Bring the flattened palm smartly straight back toward the dog's face, stopping immediately in front of the nose.
5. Drop the arm directly back to the left side.

Commands Away from the Dog

Once the team is proficient in movements with the dog in the HEEL position, progress to movements and positions with handler and dog separated by varying distances.

"End of the leash, move"

After giving this command, the handler gives the hand and voice command "STAY," then takes one step forward, right foot first, and pivots 180 degrees left to face the dog. As you make the pivot, transfer the leash from the right hand to the left. At the completion of the pivot, place the left hand in front of the belt buckle with the loop of the leash over the left thumb and the fingers curled around the leash as it continues down past the palm of the left hand.

"STAY" at EOL

When at the end of the leash with the leash in the left hand and in front of your belt buckle, give the command "STAY" (verbal and hand).

With fingers extended and together, bring the right hand to shoulder level, palm toward the dog. Push the palm toward the dog's face smartly, commanding "STAY." Smartly drop hand and arm directly to the side.

"Return to the HEEL position, move"

After you give the verbal and manual command of "STAY," step off with the right foot to the right flipping the leash to the left so that the leash rests on the right side of the dog's neck. This will keep the leash from hitting the dog in the face. Walking in a small circle around the dog to the rear returning to the dog's right side. Take up the slack in the leash and transfer it back to the right hand. Praise the dog verbally and physically.

"DOWN" at EOL

With the dog in the HEEL/SIT position, give the command "STAY" and move to the end of the leash changing the leash to the left hand. Take one step forward with the right foot and grasp the leash about 6 inches from the snap. Exerting pressure downward on the leash, verbally command "DOWN." When the dog is in the DOWN position, give the command "STAY" and bring the right foot back to the starting position. At the point when the leash pressure is no longer needed, introduce the hand gesture for DOWN. Lock the elbow, extend the fingers, and rotate the arm in a full circle to the rear, until the arm is at shoulder level and parallel with the ground, palm down. While the arm is making the circle, give the verbal command "DOWN."

"SIT" at EOL

The command "SIT" is introduced when the dog has learned the command "DOWN/STAY." With the dog in the DOWN position, the instructor gives the command "SIT DOG, COMMAND." The handler steps forward one step with the right foot, grasps the leash about 12 inches above the choke chain, exerts upward pressure on the leash and gives the command "SIT." When the dog sits, give the command "STAY," give verbal praise, then return to the original position. When the dog is sitting, without using leash pressure, introduce the manual gesture as follows. Extend the fingers of the right hand and lock the elbow. Turn the flattened palm toward the dog. Smartly lift the extended arm to the horizontal shoulder position and command "SIT." Drop the arm smartly back to the side. Praise verbally but not excessively.

Commands and Moves for the Handler/Dog at EOL
Circle dog

The handler gives the command "STAY" and steps off with the right or left foot depending on the direction of the command. As you make the circle around the dog, flip the leash around the dog's neck to the opposite side of the beginning direction of the circle. Take care during the circle movements not to stretch the leash taut causing the dog to break position.

Step over the dog

The same procedures apply as the Circle Dog, with the exception that the dog is in the DOWN position so that the handler can step over conveniently.

Straddle dog

The handler gives the command "STAY," steps forward with the right foot, lowers the leash, steps over it with the left foot, and proceeds to straddle the dog that is in the DOWN position. When the handler gets to the rear of the dog, to the left 180 degrees, step over the leash with the left foot, straddle the dog, and return to the end of the leash. As the handler makes the turn to face the dog again, he returns the leash to the left hand.

Recall dog

With the handler at the end of the leash, the instructor commands, "RECALL DOG, COMMAND." The handler gives the verbal and manual command of "HEEL" and if necessary calls the dog's name to get its attention. If the dog is reluctant to come on command, you may have to apply slight pressure on the leash with some verbal coaxing to get the dog to come. As the dog is returning, take up the slack in the leash and guide the dog into the HEEL position.

Military Drill

In all formations, the dog remains in the HEEL/SIT or marching HEEL position.

Attention

The position of attention is a two-count movement. At the preparatory command "SQUAD," the handler comes to attention. At the command "ATTENTION," the handler takes one step forward with the left foot and gives the command "HEEL." When the right foot is brought forward even with the left, the two-count movement is complete and the dog should be in the HEEL/SIT position.

Parade rest

At the preparatory command of "PARADE," the handler gives the command and manual gesture "DOWN." At the command of execution "REST," the handler gives the command and manual gesture "STAY," then steps over the dog with the left foot straddling the dog. The handler places his left hand behind his back. To resume the position of attention, use the preparatory command "SQUAD," at which time the handler gives the command "STAY." At the command of execution ("ATTENTION"), the handler steps back over the dog and gives the command "HEEL."

At ease/rest

When given the command, keep the left foot in place while the dog remains in the HEEL/SIT position.

Fall out

When given the command, the handler leaves ranks and puts the dog on break.

Fall in

The handler and dog resume their previous position in ranks at the position of attention with the dog in the HEEL/SIT position.

Right face

"RIGHT FACE" is a four-count movement. At the command of execution "FACE," the handler takes one step forward with the left foot, commands "HEEL" and pivots on the balls of both feet 90° to the right. They then take one step forward with the right foot, bringing the left foot even with the right. The handler then commands "HEEL" and returns to the position of attention.

Left face

"LEFT FACE" is a four-count movement. At the command of execution "FACE," the handler takes one pace forward with the right foot, pivots on the balls of both feet 90° to the left and commands "HEEL." They then take one step forward with the left foot, bringing the right foot even with the left and returning to the position of attention.

About face

"ABOUT FACE" is a four-count movement. At the command of execution "FACE," the handler takes one step forward with the left foot, commands "HEEL," then pivots 180° and gives the command "HEEL." On the completion of the pivot, the handler takes one step with the left foot bringing the right foot beside it, and returning to the position of attention.

Drill Formations

Four drill formations are used to teach basic obedience. Each is designed for a specific purpose, yet is flexible enough for other phases of training.

For safety, allow intervals of 15 feet between dog teams during initial obedience training. When handlers can control their dogs, you may reduce this distance.

Circle formation

In this formation, the dog can learn the HEEL position. It requires walking at the handler's side without sharp turns. The instructor is usually in the center of the circle for better observation of the dog teams.

Square formation

This formation is excellent for teaching the dog the HEEL position when the handler is making sharp turns.

Line formation

The line formation is used effectively during basic, intermediate, and advanced obedience.

Flight formation

The flight formation is introduced after the dog teams demonstrate proficiency in the circle, line, and square formations. Use it for moving groups of dog teams from one location to another.

Intermediate Obedience

This training differs from basic obedience in distance only. In intermediate obedience, use the 360-inch leash instead of the 60-inch leash. Once the 360-inch leash is attached, the handler should start at the same distance as with the 60-inch, then gradually increase distance and time spent at EOL. During intermediate obedience, if the dog fails to perform any specific command, the handler should walk back to the dog and put the dog in the desired position. While approaching, give the command "STAY," only if the dog starts to break position. After making the correction, use shorter distances for later trials. Never run back to the dog or make threatening gestures. This may make the dog break position and run.

Advanced Obedience

Advanced obedience allows the dog to learn to execute commands given at a distance, off leash. To begin off-leash training, the handler must execute basic command and movements with the dog at his side. (This gives the handler an opportunity to test the dog's reliability, and revert to using the long or short leash to correct deficiencies.) This obedience training at the handler's side should continue until the handler believes the dog will perform among other teams without hostility. As training progresses, the handler moves out in front of the dog a short distance and gradually increases the distance and time periods away from the dog. The dog's performance will determine distance from the handler. Normally, 50 feet is the maximum distance. When the handler moves back to the dog, the handler should circle around and step or jump over the dog. These movements teach the dog to remain in position until otherwise commanded. This stage of training should require only a minimum number of corrections. If the dog does not respond correctly and consistently to commands, the handler must return to the preliminary off-leash exercises and repeat them as often as required.

OBSTACLE COURSE

As an MWD team becomes proficient in basic obedience and associated tasks, introduce the obstacle course for the purpose of building the dog's confidence in negotiating similar obstacles the dog may encounter in the field. The obstacle course also conditions the dog and builds handler confidence in the dog's abilities. The determining factors for length of time spent and frequency of obstacle course use include dog's age, physical condition, and weather conditions.

Obstacle Course Training Procedures

The dog jumps or scales obstacles on the command "HUP," and when commanded, returns to the HEEL position. As in other training, first teach the dog to complete exercises on leash. This allows the handler more control while guiding the dog over obstacles. As the dog's proficiency increases, train the dog off leash. A dog may hesitate to jump over a hurdle. It is best to use a hurdle with removable boards and lower it so the dog can walk over it. Exerting pressure upward on the leash will cause the dog to balk or hesitate. When the hurdle is lowered, the team approaches it at normal speed, and the handler steps over it with the left foot and commands "HUP." If the dog balks, the handler helps it over by coaxing and repeating the command "HUP." After crossing the hurdle, the handler praises the dog, and gives the command "HEEL." As the dog progresses, add boards until attaining a height of 3 feet. Thereafter, when the handler is two paces from the hurdle, give the command "HUP." Instead of stepping over, the handler passes around to the right of the obstacle while the dog passes over it. (Allow more than two paces from the hurdle if necessary.) As the dog's front feet strike the ground, the handler commands "HEEL," adjusting the distance in front of the dog so there is room to recover from the jump and assume the HEEL position. Immediately after the dog is in the HEEL position, give praise. Vary hurdle procedures somewhat for the window, scaling wall, catwalk, and stairs. For the window, the handler must transfer the leash from the right hand to the left and throw the leash through the window catching it on the other side. If the dog hesitates, put the front feet in the window and coax the dog through. For scaling the wall, the dog must have more speed on approaching and you must give the "HUP" command sooner. Adjust the wall to the dog's abilities during initial training, gradually increasing the incline. For the catwalk, the handler may have to guide the dog onto it and steady the dog's balance while it crosses. The dog must walk up and down the stairs. If wet, remove the water from the stairs prior to use. The handler may have to walk over the steps with the dog if it hesitates.

CONTROLLED AGGRESSION

With exception of detection training, controlled aggression is the most intricate aspect of military dog training. Supervisors must ensure that each dog is trained and maintained at maximum proficiency.

Attack "GET 'EM"
Give the command only once. Give further encouragement if necessary. During on leash agitation, the handler must maintain position and balance by spreading the feet at least shoulder-width apart, one foot slightly forward of the other. Flex the knees and bend slightly at the waist. While extending the arms, unlock elbows. Not following this procedure could cause the handler to lose balance and cause serious injury to another handler or dog.

"HOLD 'EM"

Give command in an encouraging tone of voice while the dog is biting. If the dog releases the bite, repeat the command "GET 'EM," then
repeat "HOLD 'EM."

"OUT"

Give this command to cue the MWD to cease attack. A properly trained
dog will release the bite and on receiving the command "HEEL" return
to the handler. Upon successful completion, the handler must physically and verbally praise the dog. If the dog does not release the bite,
the handler should wait 3 seconds and repeat the "OUT" command
or command "NO-OUT." If the dog does not release after the second
command, the handler should repeat "NO-OUT" and apply a physical correction. **NOTE:** The dog must know the task before using the
physical correction! You can teach the "OUT" command before the dog
is actually biting the wrap. Do this with EOL agitation. The handler
should give the command "OUT, HEEL" and use the leash to guide
the dog back to the HEEL position. The handler must physically and
verbally praise the dog when it ceases aggression.

"STAY"

A properly trained MWD will remain in the stay position until you give
another command. During controlled aggression exercises, use the command "STAY" to notify the agitator that you are ready for exercise ini-
tiation. You may find the DOWN position helpful in preventing some
dogs from breaking position.

"WATCH 'EM"

Given in a very suspicious tone of voice to put the MWD on guard. If
during agitation, the dog loses interest, repeat the command.

AGITATION

Agitator's Role

The agitator plays an important role in agitation exercises; therefore,
thoroughly instruct persons acting as agitators on what to do. As an agitator, you may use a supple switch, a burlap bag, an arm protector, or a

rag to provoke the dog without actually striking it. The dog's level of aggression will determine the need for using such training aids. The agitator's actions should replicate actions of real-life subjects the dog may encounter.

Aggressiveness

To determine the degree of aggressiveness or develop aggressiveness of the dog, conceal the agitator upwind of the dog team. The handler, while maintaining a safety leash, approaches the area concealing the agitator. The agitator will attempt to attract the dog's attention through normal suspect/intruder actions. Weaker dogs may require the agitator to slightly increase movements and/or make additional noise to gain the dog's attention. Meanwhile, the handler must watch the dog closely to provide timely assistance by encouraging the dog in a low suspicious voice, to "WATCH 'EM." When the dog detects the intruder, the handler must encourage the dog immediately. If the dog shows no interest, the agitator should show himself and move away suspiciously as the team gets within 10 feet.

Under aggressive

This type of dog will fail to exhibit interest in the agitator even as he moves away suspiciously. To develop aggression in these dogs, use the chase method. The agitator provokes the dog. As the dog shows aggression, the intruder will run away while continuing to make noise, while the team gives chase. After running 20 yards or so, the agitator will throw up an arm to indicate the direction they intend to turn. The agitator will turn in that direction and the team will turn in the opposite direction. The handler should exercise care not to jerk the dog off the chase, causing an unintentional correction.

CONTROL

Building Control

To build control, give the dog "STAY" in the HEEL/SIT position and have the agitator move in from a distance of approximately 20 feet. The agitator may use a play rag, puppy tug, or arm protector. The agitator should approach the dog team in a manner that arouses the dog's suspicion. The handler should give the dog the "STAY" command and reinforce the command as necessary. The agitator will then retreat back to the starting position and cease movement. The handler should physically and verbally praise the dog. If the dog breaks position, the handler should command "NO-STAY," guide the dog back into position and repeat the command "STAY." If the dog continues to fail the "STAY" command, the handler must adjust the severity of the corrections to meet the level that will effectively change the dog's behavior. Once the dog is proficient in this scenario, the agitator will move in closer to the dog team and act in a more suspicious manner, thus increasing dog stimulus to aggress. As the dog becomes proficient at this level, introduce the wrap and command the dog to bite. Use the same process to train the dog to release the wrap as you did in the initial scenario. You may need to increase the level of correction or revert to the previous method. If the dog fails to progress at this level, return to the initial scenario.

Commands of "OUT" or "NO-OUT"
After the dog demonstrates proficiency in biting and holding, the agitator can hold the rag/protector. The handler will command "OUT" or "NO-OUT" when the agitator ceases movement.

Hurt or bitten intruder
If for any reason the intruder is hurt or bitten, he should signal the handler by raising the free arm above his head. The handler should immediately give the "OUT" command and physically gain control of the dog.

False Run (MWDs Trained Under the Out and Guard Method— Field Interview)
The field interview is a practical replacement for the traditional false run and designed to demonstrate the MWD's ability to be tolerant of nonaggressive movements or situations dog teams can be exposed to throughout their tour of duty. Through this training exercise it enables the handler to gain complete control over the MWD while subjected to various non-suspicious actions by a subject. As the dog's proficiency increases, this exercise will be conducted off leash.

Training procedure

Put the dog in the HEEL/SIT or DOWN position and give the "STAY" command. The subject, wearing the arm protector, starts at a distance anywhere from 15 to 100 feet from the dog team. MWD proficiency levels of each dog trained in this exercise scenario will determine the start distance for this portion of the exercise. Once the distance has been determined, the subject will start off by facing the dog team in a non-suspicious manner and will walk toward the team by taking an indirect route. The purpose of this type of movement is designed to simulate an individual who upon initial contact or observation is nonthreatening to the dog team, others, or resources. As the subject reaches the dog team's location, the subject reaches out (using the wrap-protected hand first) and simulates shaking the handler's hand and engages the handler in a normal conversational tone of voice. After the nonthreatening verbal interaction between handler and subject, the subject simply turns and walks away from the team. Again, the subject's intension/movements are nonthreatening toward the dog team as he departs. Throughout all stages of this exercise, the dog must remain in the HEEL/SIT or DOWN position and should not show any signs of aggressive behavior toward the handler or subject. Handlers and trainers should use caution not to misinterpret the dog's natural behavior to be interested or curious of the person, as he moves closer toward the team as aggressive behavior. Some dogs may attempt to break the HEEL position as the subject approaches just because they are simply interested in a nonthreatening sense of the approaching individual. Should a dog show this type of behavior and it is determined as non-aggressive behavior, the handler must reinforce the HEEL command, ensuring the dog returns and stays in the HEEL position. If the dog stays and doesn't exhibit aggressiveness, give the dog lavish praise. If the dog breaks position, correct it immediately and repeat the exercise. During repeat exercises, as the dog shows progress the subject should now take a more direct route toward the dog team. Take care to let the dog have a bite at irregular intervals to keep the dog from becoming too frustrated, and as an indirect form of praise. Correction must not be too harsh. Training in "STAND OFF" and "FALSE RUN" phases can be very difficult, especially for dogs with a high fight drive. Dogs demonstrating a high proficiency rate for this particular exercise will naturally have the distances and time ratios of dog team/subject interaction varied. Dogs with a lower proficiency rate for this exercise will need specific distances and time ratios closely monitor and controlled to build the MWD's proficiency level for this exercise.

Pursuit and Apprehension

Pursuit and apprehension are used to teach the dog, on command, to pursue, bite, and hold an individual.

Training procedure

Proficiency in all phases of obedience and timely response to commands is required prior to starting controlled aggression training. Begin with the MWD in the HEEL position off leash and give the command "STAY." To begin the exercise, the subject should stand or move around suspiciously at a distance of 40 or 50 feet. Prior to releasing the dog, the handler will give a warning order *"Halt or I will release my dog"* and warn bystanders to cease all movement. (NOTE: Refer to your unit SF operating instructions for exact verbal challenging procedures/instructions.) When the handler commands "GET 'EM," the dog should pursue, bite, and hold the subject until commanded to release from the bite. The handler follows the dog as closely as possible. If the subject stops or indicates surrender, the handler will immediately call the dog off the pursuit (standoff). If the dog makes contact with the subject, the handler will call the dog "OUT" once the situation is under control. DURING TRAINING ONLY, when the dog is biting, the handler provides encouragement and commands "HOLD 'EM." After a short struggle, the subject ceases movement, and the handler commands "OUT." Give praise when the dog returns to the HEEL position. There may be times when the dog will release from the bite but will hesitate in returning to the handler. Should this be the situation, the handler will use verbal encouragement to refocus the dog's attention back to the handler. As the handler gains the dog's attention and the dog is returning to the HEEL position the handler should cease verbal encouragement/praise until the dog is in the HEEL position

Training realism

Conduct training in the dog's working environment when possible. Training problems must replicate "real-life" scenarios as much as possible to include the frequent use of hidden arm protectors.

Search of a Suspect

Search apprehended personnel as soon as possible. In most instances, it is best to have another security forces person conduct the search with the dog team as backup. If no other police personnel are present, the handler may search the suspect with the dog in the GUARD position. Ensure the dog can observe the agitator/suspect at all times.

Training procedure

The handler will position the suspect 6 to 8 feet in front of the dog, facing away from the dog. Prior to the search, place the dog in either the SIT or DOWN position and inform the agitator/suspect not to make any sudden or aggressive movements or the dog will attack. The handler gives the dog "STAY," moves forward (right foot first) to search the agitator. Do not pass between the agitator/suspect and the dog. After searching both sides, the handler positions him directly behind the suspect. If the dog attempts to bite again or shows undue interest in the agitator/suspect, issue an immediate correction. When the dog returns to the proper HEEL position, give lavish praise.

Re-Attack

During a search, the MWD must learn to re-attack. If, during the search, the agitator/suspect attempts to run away or attack the handler, the dog must immediately pursue and bite and hold the agitator without command. In the early stages of or periodically during proficiency training, the handler may have to command "GET 'EM." Excessive training in this area may result in a dog anticipating the moves of the agitator/suspect, causing loss of control by the handler.

Escort

After apprehending and searching a suspect, you may find it necessary to escort the apprehended individual out of the immediate area to a vehicle. After recalling the dog to the heel position, directly behind the suspect, the handler takes control of the suspect by placing his hands on the suspect's shoulder and escorting the suspect. The dog may heel on the handler or slightly forward on the side of the decoy to ensure an effective escort.

Standoff

The purpose of the standoff is to develop control needed by the handler to call the MWD back from a bite and hold command.

Training procedure

The agitator moves toward the dog acting suspiciously. At a distance of 4 feet, the agitator turns and runs. When the agitator gets about 30 feet from the team, the handler commands "GET 'EM." When the agitator hears the command, he should stop and cease movement. The handler commands "OUT" and, if necessary, "NO-OUT." After the dog OUTs, the dog must SIT, DOWN, or STAND within the immediate area of the agitator.

Gunfire and Cover Command

The primary purpose of gunfire training is to condition the dog to perform all required tasks satisfactorily when gunfire is introduced to the scenario. The dog should not react aggressively, unless commanded by the handler, nor should it display an avoidance behavior toward gunfire. Training should be conducted both with the handler and agitator/suspect firing the weapon individually and together. Depending on the various stages of training, however, gunfire associated with agitation or bite training should be kept to an absolute minimum, and then only to determine if the dog still performs satisfactorily in the presence of gunfire.

Training procedure

Under no circumstances will a dog be backed down or defeated in gunfire training. Conduct gunfire training using only authorized blank ammunition. Ensure the muzzle of the firearm is always pointed in a safe direction.

Conduct gunfire training

It is best to use a small caliber weapon, casually and intermittently. Begin initial gunfire training from a distance of at least 70 yards and include it in all phases of training. As the dog performs satisfactorily, move gunfire gradually closer to the dog. Reward the dog when it ignores gunfire. Do not reward the dog when it shies away from or aggresses toward gunfire.

As the dog accepts gunfire at varying distances, introduce advanced training. Progress to larger caliber weapons and, if possible, expose the dog gradually to mortar, artillery, and grenade simulators. When the dog is proficient in gunfire, introduce the command "COVER." This simply means that on the command of execution, the handler gives his dog "DOWN" and assumes the prone position. Great care and caution must be used to prevent a potential safety mishap during this phase of gunfire training. Training supervisors must understand during this phase of training the handler will be placed in a difficult position to protect himself (from a dog attack) while in the prone position next to his dog. As in all initial stages of dog training, training supervisors and handlers need to remember there might be a period where the dog may not produce the desired behavior at first; therefore, practice and patience are musts for the dog to master this task.

SCOUTING

Scouting is the most effective procedure to locate intruder(s) hidden in a large area. The following factors affect the MWD's ability to scout.

Wind

Wind is the most important and variable factor in scouting. It carries the scent either to or away from the dog; therefore, the handler must remain aware of direction and velocity at all times and must fully understand how this affects the dog's ability to successfully perform this task. Handlers must be capable of accurately identifying wind direction in all types of terrain and/or situations without outside sources. For example, at night or other limited visibility situations it may not be feasible to drop hair or a blade of grass to check wind direction. The best way to check the wind direction is to remove head gear and turn slowly until the breeze creates a cool feeling on the upper forehead. As mentioned earlier, not only is it important to know the proper wind direction it is equally important to know how wind speed will affect your scout. An ideal wind speed is difficult to pinpoint, but normally slight, steady, and consistent wind conditions will produce the best scouting results for dog teams.

Terrain

The next important consideration is terrain. Besides man-made structures, there are trees, bushes, large rocks, high grass, and many other natural variations. Odor cannot pass through obstacles, so it must go over, under, or around them. Handlers must understand how these obstacles come into play when performing scouts with their MWDs.

Additional Factors

Additionally, the MWD trainer and agitator must remain aware of the wind direction and their route when walking within the area that the dog will search. Agitators should always approach the area from the upwind flank to ensure the dog does not cross the path and track the agitator. Trainers should also take the same precaution when training multiple dog teams. The agitator should be moved between scouting problems so dogs cannot track each other. Other factors that affect scouting abilities and conditions are rain, snow, sleet, temperature, and humidity. These factors affect the odor concentration or scent cone in numerous ways.

SCOUTING PROBLEMS

Set your scouting problems to match the proficiency of the dog team. For example, when quartering a field, an advanced dog may be able to consistently locate the decoy from 40 yards. This should be considered when bounding forward when quartering the terrain to locate decoy. Use chase agitation to build drive in weak dogs. Once teams are proficient in initial scouting problems, advance to realistic problems including vast areas of the installation. Vary the terrain to include wooded as well as developed areas. Use your imagination and set real-world scenarios, including the use of other flight members for backup and response forces. There should be an even balance of bite training incorporated into scouts.

MAINTAINING PROFICIENCY

When a team arrives at a proficient level to scout and clear an area, there are many ways to keep the dog team proficient.

Field Problems

Designed to evaluate use of scouting principles. The area should have a variety of terrain features, and the handler must know the area boundaries.

Patrolling Exercises

Usually consist of point-to-point posts; however, a specific or a designated area might need securing.

Training procedure

The trainer places several intruders along the line of patrol 75 to 100 yards apart along the route the dog team takes. Position the agitators off the line of patrol, far enough to challenge the dog's detection capabilities, but not defeat it. Set up all three types of responses to include sight, scent, and sound. At the conclusion of the exercise, the handler indicates the number of agitators found.

SECURITY PROBLEMS

Set up realistic problems with the goal of extending the period of time the dog team remains alert on regular sentry posts. Supervisory personnel can use these problems to best evaluate the dog's training and the abilities of the handler to control the dog.

Alternate Teams

Alternate teams between different types of posts as training progresses. Initially, each team is used on-post for about 30 minutes before the agitator hides on the post or tries to penetrate the post.

Advanced Training State

At this advanced stage of training, do not use the command "FIND HIM" to get the dog to respond unless it is absolutely necessary. Once the dog responds, replicate normal apprehension and escort procedures.

Extended Training Time

After a few nights of this training, the team's tour of duty is extended to either 4 or 6 hours, as determined by posts and training time. The extended training time is necessary to condition the dog to remain alert and watchful over a normal tour of duty. Vary the number of penetrations for each team in time and number. This variation keeps the dog alert for penetrations.

Penetrations

Penetrations serve two purposes: to check the security of an area, and to maintain a patrol dog team's proficiency. The penetrator tries to enter the post undetected and, if successful, hides along the handler's route where he must allow the dog to detect the intruder.

Patrol Dog Team's Training Benefit

A patrol dog team gains no training benefit from an exercise in which the agitator/intruder penetrates a post with the intent to elude detection. The penetrator must not use the same route or time of approach. If he does, the handler and dog begin to anticipate arrival. The penetrator must use cover and concealment when penetrating a post, to avoid revealing the position before reaching the post perimeter.

Detection Capabilities

Training emphasis is placed on developing the detection capabilities of the dog. Sometimes it is necessary for the penetrator to make his presence on the post more obvious.

Diversionary Tactics

During the early stage of training, the penetrator must not use diversionary tactics because these tactics may confuse the inexperienced dog team.

Team Proficiency Training

An effective penetrator must have the dog team's proficiency training in mind. He must employ sound judgment and adapt methods to the situation matched to the proficiency level of the team. These practices apply during training as well as under field conditions.

BUILDING SEARCH

Use a building search to locate an intruder hiding in a structure.

Factors Affecting Building Searches

Factors that influence an MWD's ability to scout also affect its ability to locate an intruder inside a building. A variety of air currents are common inside buildings just as they are outside buildings.

1. Wind direction outside buildings correlates with the direction of air currents inside by filtering through any openings such as windows, doors, vents, and cracks in floors.
2. Type and size of buildings and wind direction will affect the dog's ability to detect an intruder.
3. Air conditioning units, fans, and heater blowers affect the speed and direction of airflow. Changing air currents can confuse the dog in its effort to locate the intruder.
4. Temperature inside and outside a building may influence the concentration of odor. Cold temperatures will keep the odor closer to the surface, while warm temperatures will cause the odor to rise.
5. Residual odor from personnel who recently departed the building may serve to distract the dog.

BUILDING SEARCH TRAINING

Set your building search problems to match the proficiency of the dog team. Use chase agitation to build drive in weak dogs. Conduct initial training with a 6-foot or a 30-foot leash. Make your problems more difficult as the dog progresses. Once your team is proficient in initial building search, conduct advanced searches in realistic environments. Use your imagination to set problems that challenge the capability of the team. Use other flight members to act as backup, over watch, and response forces. As in all training scenarios, the safety of the handler, agitator, and dog must be at the forefront of your scenario planning at all times.

Initial Building Search Training Procedures

In initial building search training, allow the dog to see, hear, and smell the intruder just inside the building at the entrance door. Gradually move the intruder into a room, allowing the dog to detect by the use of intruder movement (vision and sound); then progress to the dog locating the intruder by odor. When the dog responds to the odor of the intruder, by barking, scratching, etc., ask the dog "WHAT YA GOT?" Verbally reward the dog as it makes the required response and enter the room to allow the dog to bite the intruder. As the dog responds correctly to the following trials, move the intruder into the next room to teach the dog the intruder location was moved. Gradually lengthen the search, one room at a time until the dog searches the entire building. Then randomly position the intruder throughout the building. Conduct all building searches in a systematic manner, preventing duplication. The handler should always clear an area or room before passing it and maintain an avenue of escape. Dogs are trained to make one of the following final responses upon locating an intruder: vocal (bark, growl, whine), scratching, biting, or ceasing movement at the intruder location.

Intermediate Building Search Training Procedures (On Leash)

Hide an intruder in the building for a designated period of time prior to the search. The trainer/supervisor should adjust the time based on dog's ability, building size, and difficulty of search. The handler should cue the dog to start searching for the intruder with the command of "FIND 'EM" to begin a systematic search at the appropriate starting point. The handler allows the dog to clear the building observing any indication the dog detected the intruder. Tell the handler the location of the intruder during training to help identify responses. The trainer should accompany the team occasionally to give added advice and assistance as needed. When the handler is sure the dog has detected the intruder, reward the dog with a CRS using verbal and physical praise. You may use a bite reward if it enhances the dog's proficiency. Remember, the dog is also operating on a VIRS at this point. To assure the dog that the intruder is really present, you may find it necessary to reveal the intruders presence by noise or even show a portion of the intruder's body. Eventually the dog develops proficiency on odor cues only. When possible, terminate all building search exercises by having the team escort the intruder from the building.

Advanced Building Search Training Procedures (Off Leash)

Cue the dog to start a systematic search similar to the on-leash search. The handler must follow the dog as much as possible to keep it in view. The handler must react instantly when the dog responds. Conceal the intruder in a location not accessible to the dog. The intruder should remain quiet and allow the dog sufficient time to search out the hiding place. A dog that has performed well to this point will have learned to search systematically and efficiently on its own without the handler close by. A dog should eventually search out the intruder without assistance from the handler. If the dog responds on an area where the intruder was previously hidden, make certain the area is cleared, then repeat the command "FIND 'EM" and continue the search. During actual (not training) searches, a single trial where no reward is given constitutes an extinction trial. This single trial will not degrade the dog's subsequent performance due to the fact the dog is on a VIRS. Once you have trained a dog on this schedule, you have proven behavior is highly resistant to extinction.

Actual Building Search Procedures

When conducting an actual building search, several factors must be considered:

1. Potential danger to the handler and MWD.
2. Type and size of building (speed of search).
3. Time of day or night.
4. Evidence of forced entry.
5. Known or suspected contents of the building.
6. Possibility of innocent persons inside.

- After considering the advice of the handler, the on-scene commander will determine whether to search the building on or off leash. The handler must announce in a clear, loud voice that he will release the dog to search the building if no one appears within a specified time. This allows intruders the opportunity to surrender or innocent persons the opportunity to make their presence known. If no one appears, the handler will allow the dog enough time to clear the immediate area before proceeding. The handler will then follow the dog to each unclear room until the building is cleared.

- If an intruder is located, recall the dog, challenge, and apprehend the intruder. Another security member should accompany dog teams. This individual should follow at a discreet distance to avoid interference with the search. This additional person is needed to provide over watch protection to the handler because the handler must focus his attention on the dog and not necessarily the threat associated with the building search itself. If an intruder is found, the security forces person who accompanied the dog handler can perform a quick body search and remove the individual.

- Do not enter the building until backup units have secured all avenues of escape. If forced entry is indicated but search results are negative, consider using the dog's scouting or tracking capabilities. If the situation allows, the building custodian should be called to the scene prior to security forces members entering the facility. By having the custodian on scene, he should be able to provide more in-depth information of the facility and possibly identify what is out of place and what is not associated with the building.

TRACKING

Some MWDs are completely unsuited for tracking and show no willingness to track. Nothing can be gained by continually trying to make one of these dogs track. Therefore, once a kennel master or trainer is able to document a dog's inability to track, further training in this task may be stopped. Dogs that demonstrate a definite ability to track must remain proficient by consistent training.

Tracking Dogs

Tracking dogs are utilized during combat to locate enemy by the scent they leave on the ground. In law enforcement, tracking dogs are often used to search for fleeing felons, lost/missing persons, and evidence. Because MWDs are not trained to track during initial training at Lackland AFB, it is left to kennel masters to identify dogs with tracking potential. Therefore, it is important to have general knowledge and understanding of the tracking dog capabilities.

Conditions Affecting Dog Performance

Before beginning training with your dog, you must understand some of the conditions that affect your dog's performance: wind, surface, temperature, distractions, and age of the track.

Wind

The dog takes the human scent not only from the ground, but also from the air near the ground. A strong wind can spread the scent causing the dog difficulty in detecting the scent. A strong wind may also cause a dog to depend upon its scouting ability (track laid into the wind) to find the tracklayer instead of tracking. A wind blowing across a track (track laid crosswind) may cause the dog to work a few feet downwind of the track. To encourage the dog to pick up the scent directly from the ground, all initial tracks should be laid downwind from the starting point. Once the dog becomes proficient, use tracks that combine different wind conditions.

Surface

The ideal surface for tracking is an open field with short, damp vegetation. A hard dry surface does not hold a scent well. Heavy rain can dissipate or mask the scent. In contrast, a damp surface will allow the scent to remain.

Temperature

The scent dissipates faster when the temperature is high. The early morning or late afternoon hours are more favorable tracking periods. Rain will quickly dissipate the scent.

Distractions

Some odors can mask the human scent the dog is following. Conflicting scents, animal odors, smoke fumes, chemicals, and fertilizers affect the dog's ability to detect and follow a track.

Age

The age of the track is another factor that must be taken into consideration. It is more difficult for the dog to follow an older track.

Types of Tracks

For training purposes there are three types of tracks: initial, intermediate, and advanced

Initial track

The initial track is laid downwind and runs from one point to another for about 50 paces. The tracklayer leaves a scent pad approximately three square feet by stepping in the entire section. After laying a scent pad, the tracklayer takes short steps closely together in a straight line downwind to the end of the track. This will place more scent on the track making it easier for the dog to learn the task. During initial tracking, encourage the dog to watch the tracklayer. Preplan the initial track so all personnel involved know the start and end points. The handler must know the exact location of the track. For all phases of tracking, establish records to document conditions. Furnish the handler a precise map and detail the length and age of each track. The area used for an initial track should have low vegetation (i.e., mown grass). There should be ample room to lay the track. Keep distractions to a minimum.

- *Initial Track Procedures.* After the tracklayer finishes laying the track, have the dog team approach and stop 6 feet from the scent pad. During initial training, attach the 360 leash to the choke chain to control the speed and assist in keeping the dog on the track. You may use a collar or harness. At the scent pad, coax the dog to smell the scent pad, command the dog "TRACK," pronouncing it in a slow, drawn-out, pleasant manner: "T-R-A-A-A-A-CK." Casting is given by making a sweeping downward and out motion with the palm of the hand up. Keeping on the track, give the dog half leash and move along the track. Whenever the dog strays from the track, slow the pace until the dog recovers or returns to the track. If the dog strays off track, stop, call the dog back, have it smell the track, and repeat the command, casting the dog out to half leash. To encourage the dog to track, use a mild form of enticement or small pieces of food at the scent pad and throughout the track. Allow the dog to track at a slow pace.

Place several pieces of food to indicate the end of the short track. After the dog consumes the food, praise the dog lavishly as you exit the area. This may take several trials before moving on to the next phase.

Intermediate track

The intermediate track includes turns the dog must follow and articles that must be found. Articles are small pieces of wool, leather, rubber, or cloth. As in any track, preplanning is a must so the handler can observe and assist the dog during turns and in locating articles.

• *Intermediate Track Procedures.* The starting scent pad is smaller than the scent pad used for the initial track. The length of the track is 100 paces and will include two 45-degree turns. Articles are placed to reinforce the tracklayers scent on the trail. They are also used to indicate change of direction in the track. The scent is placed on the articles by rubbing them between the hands. The dog is not required to pick up the articles, but should indicate the exact location. The dog should respond by lying down with the article directly in front of it. The response on the article serves as the point to reward the dog and a resting place. Give verbal praise and a small piece of food after the dog downs at the article (no more food is placed on the track itself). It must be emphasized that incentives are extremely important in getting the dog to follow a scent. The article can be used as a refresher scent if the dog loses the track. The down indication of the articles should be taught separately to avoid any pressure on the track. To create a new problem for the dog, lay the track crosswind. The dog may work a few feet downwind from the track to pick up the scent. Initially, the track should be fresh and increased in length and age as the proficiency of the dog increases. After making a scent pad, the tracklayer walks normally along a preplanned route, occasionally shortening the pace to put down more scent. Wherever an article is placed, the tracklayer stops and stands in that area to increase the scent. Going into and coming out of a turn, slowing the pace makes small scent pads.

Advanced track

Dogs that have shown a marked degree of proficiency in tracking will use the advanced track. Not every dog has the ability to track for long periods or follow old tracks. Notice that scented articles are used, the turns are sharper, and a diversionary track is used.

- *Advanced Track Procedures.* The track should be approximately 1–2 hours old and about 1 mile long. The tracklayer lays a track by making a scent pad and then walking at a normal pace but occasionally breaking into a run. Increase turns to 90 degrees and include downwind, crosswind, and upwind scenarios. Use a diversionary track to teach your dog the difference between the primary and a cross track. To prevent confusing the dog, have a diversionary tracklayer quickly cross the primary track. Thus, to observe your dog's reaction, you must know exactly where the tracks cross. At every other turn, the tracklayer makes a scent pad if the trainer feels the dog still needs assistance. After the turns, place articles as a reward for accomplishing the turns. If articles are not used at the turns, make separate scent pads for the dog to follow. In planning this track, prepare a detailed map so the handler can help the dog when necessary. To ensure a successful performance, the handler and trainer must stay flexible in their approach to tracking and allow latitude to make necessary adaptations.

Regaining lost tracks
As the dog advances through the track, it may lose the scent and wander off track. The key handler action is to recognize the dog has lost the track and stop immediately. When the dog loses the track, the handler should recognize a change in the dog's mannerisms. The dog may regain the scent on its own. If your dog does not regain the track by itself, take it back to the last known location of the track. Command "TRACK" and follow your dog along the track. If this does not work, use a more advanced method to regain the track. These methods include the spiral, cloverleaf, and figure eight patterns. These methods involve walking your dog in a pattern so it has an opportunity to regain the track.

- Maintain tracking proficiency by performing at least one advanced track per week. Set up the tracks to exercise and reinforce the dog's capabilities and provide enough variety so the dog does not learn to anticipate the route of the track.

DECOY TECHNIQUES

The decoy (or helper) plays a vital role in developing the drives of an MWD. Trainers, handlers, and decoys should know the dog's temperament and gear the training to build a solid balance of prey and defense drive. Decoys must always remember their ultimate goal is assisting in the building of the dog's proficiency levels and, aside from the safety of themselves and other personnel involved in dog training, this should always be at the forefront of training.

Temperament
Temperament is the combination of all of a dog's mental and emotional attributes, disposition, and personality. By understanding and evaluating temperament, we can predict trainability in any working dog. Experienced trainers can modify behavior and cover temperament flaws; however, you cannot completely change basic temperament.

Instinct
Instinct is a dog's innate response to certain stimuli, independent of any thought process such as chewing, vocalizing, digging, leg lifting, and scratching. Instincts most often have their roots in survival or reproduction.

Drives
MWD drives is discussed later in chapter 4.

Imprinting
Imprinting is an initial impression on a dog that will evoke a lasting or permanent reaction or behavior, usually associated with the untrained dog's initial learned reaction to a given stimulus or set of stimuli.

Compulsion
Compulsion is using the application of pain or negative stimuli to extinct a behavior, evoke a response, or otherwise modify a dog's behavior. Use of compulsion in patrol dog training can adversely affect drives and ultimately result in undesired behaviors. One example includes a dog that avoids the handler after a bite. Through excessive use of compulsion,

the dog has associated the handler with negative stimuli (correction) and therefore avoids the handler to delay or avoid the correction. Another example of the inauspicious effect of compulsion is a dog that slows its pursuit as it reaches the decoy and doesn't fully commit to the bite. In this case the dog has associated the decoy or the distance with a previous correction in "standoff" training.

Anthropomorphism
Anthropomorphism is to place human characteristics, motives, or emotions on a dog. Example: "My dog is bored with training and will not work." This type of statement contradicts quality training and exemplifies a misunderstanding of the animal.

Shaping
Shaping is rewarding nearly correct responses as a dog is learning a task. As the training continues, reward only those responses that are more like the desired final response. Shaping is highly effective in teaching a task without compulsion.

Rewards
Rewards are born out of drives and used to evoke the desired behavior. The anticipation for the reward drives the dog more than the reward itself. The drive for the reward can help trainers predict trainability. Use the bite, slip, and carry as a reward to satisfy prey, defense, and fight drives.

Learning Curves
Learning curves is an analytical theory that depicts fluctuations in an animal's ability to learn a task. The curve will depict the starting point, the peak, the drop off, and the flat areas of a dog's learning abilities. As the trainer begins to teach a task, the dog is eager to satisfy the drive—the starting point of the learning curve. As training progresses, the trainer can apply the learning curve using care to terminate the training session at the peak. Training beyond the peak will push the dog into the flat area. The dog is unable to properly learn a task in the flat area and will invariably respond incorrectly. At this point the trainer is counteracting the positive learning that occurred in the beginning of the trial. The

ideal training session will flow through the incline of the curve and stop at or near the peak. By consistently following this routine, the trainer ensures that all training is conducted in a positive manner, ingraining an enduring desire for the dog to successfully complete the tasks. Failure to understand and track learning curves is common among dog trainers and will lead them to assume that the animal is unable or unwilling to comprehend and complete a task.

Presenting for the High Bite (Decoy)

As the dog is in pursuit, the decoy waves the sleeve at shoulder height, ensuring that the dog targets high. As the dog leaps for the bite, the decoy pulls the sleeve in at chest level and forces the dog to fully commit to the bite. This method instills confidence and fight drive in the dog. This will transfer to the actual (suspect) bite. Failure to build this confidence will result in the dog failing to commit to the actual bite because it has anticipated the presentation of the sleeve.

Working the High Bite

The decoy works the sleeve at chest level while standing in an upright position. This forces the dog to give a harder and fuller bite. While working the dog, the decoy gives as the dog attempts to bite deeper, and pulls as the dog lets up. This conditions the dog to maintain a hard and full bite. A dog that is trained with the high bite is less likely to nip a suspect's clothing and more likely to fully bite and hold the individual.

Pull Down in the Bite

If the bite is weak or mouthy, the handler and decoy work together to build the dog's bite. With the dog on a 6-foot leash and leather collar, the handler applies backward and downward pressure on the leash while the dog is on the sleeve. The decoy works the dog in the high bite. The handler and decoy must apply pressure when the bite is weak causing the dog to fight harder for the full mouth bite. Then simultaneously release the pressure and allow the dog to readjust the bite. Do this two or three times to build the dog's bite. As the dog takes a fuller and stronger bite, reward the behavior by slipping the sleeve and allowing the dog to carry or parade with it in a circle.

Confidence Bite

The confidence bite is used as a stress relief for the animal. Because the presence of a decoy causes stress in the dog, you can use this technique to release the stress before a training session. Simply allow the dog to attack the decoy and take the sleeve. Let the dog carry the sleeve in a wide circle at a medium gait. Do not let the dog thrash or manipulate the sleeve, and do not use verbal or physical corrections with the dog. Once the dog has calmed down, the decoy can entice it with a second sleeve.

Reward Bite Method

The reward bite develops a willingness to release the sleeve and return to the handler by using positive motivation. This method has helped solve several long-standing problems in MWD training, including failure to release a bite, attacking during a standoff, hesitation in the attack, handler avoidance, and handler aggression. The reward bite is divided into three progressive steps, all of which are conducted with the dog on a 30-foot leash.

Double ball method (Step 1)

This technique teaches a willingness to release while the animal is in a low state of drive. It also produces willingness for the dog to return to the trainer. The trainer uses two identical prey reward items such as balls, Kongs, tug toys, jute rolls, or play rags. The trainer throws one of the items and the dog is sent to retrieve it. Once the dog returns to the trainer, it is enticed with the second item. This creates a conflict within the animal between the desired item and that which it already possesses. The conflict leads to a willingness to release the possessed item, which has no movement, and rewarded with the desired item. The trainer must add enough movement to the second item to build the desire to release. Do not reach for the item that the dog possesses, let it drop to the ground, and immediately reward the dog with the second item. To end this training session, provoke the release and escape the dog away from the area. Do not progress to the next step until the dog willingly releases the prey item.

Double decoy method (Step 2)

This technique follows the same principle as the double ball method, releasing the dead object (no movement) for the one with life (movement). Adding the presence of the decoy in this step evokes aggression in the form of fight drive and makes it more difficult for the animal. Step 2 comprises three phases.

- *Phase I.* The dog is sent for a bite and the first decoy slips the sleeve. The trainer should then walk the dog in a circle. Do not allow the dog to drag, thrash, or manipulate the sleeve; these behaviors degrade the training. To prevent this, increase the gait causing the dog to lift the sleeve high for the carry. The second decoy evokes the dog by adding life to the other sleeve. The dog's focus will shift from the dead sleeve to the one with life. The trainer should not correct the dog or interfere. Let the natural drive dictate the dog's behavior! The dog will release and transfer to the sleeve with life. The second decoy rewards the dog with the slip of the sleeve. End your session by evoking a release and escape the dog away from the area. Once the dog shows consistency on this phase, progress to Phase II.

- *Phase II.* Begin as in Phase I by sending the dog for a bite on the first decoy while the second decoy is out of the dog's view. The first decoy will not slip the sleeve, after the bite he must freeze and offer no fight or life. The second decoy is introduced and moves close to the dog. Once in position he provides life in the second sleeve, this stimulates the dog's prey drive. The dog will release the first sleeve and take the one with life. There is a period of conflict at this point, which will vary from dog to dog. Allow the dog time to make the transition! The handler should not interfere. The key is to let the dog accomplish the tasks through positive reinforcement. Verbal or physical corrections will void this training process and detract from the dogs learning ability. When the dog transfers, the second decoy slips the sleeve providing the reward. Let the dog carry the sleeve in a circle as in Phase I.

- *Phase III.* Once the dog consistently releases the sleeve in Phase II, increase the distance between the two decoys, and add the verbal cue "OUT." Give the cue in a moderate tone of voice and time it with the dog's natural release. At this point the trainer should know the dog well enough to predict the release. Use successive approximation to add distance between the two decoys, allowing the dog to run back and forth for each bite.

Locate the trainer between the decoys and encourage the dog as it makes the transfer. The goal in Phase III is to have your trainer, dog, and decoy positioned as in a standard controlled aggression training session. The second decoy is located behind the dog team out of sight. As the first decoy freezes, the handler commands "OUT," the dog should release cleanly and return to the handler with enthusiasm because it anticipates the reward bite. As the dog is returning to the handler, it is redirected to the second decoy. From this point begin to extinct the second decoy by commanding the dog to "HEEL"; follow the proper heel with a reward bite on the same decoy. As in all aspects of dog training, vary the routine to prevent anticipation on the dog's part.

Obedience bite

This training principle is an extension of the reward bite; it consists of the same techniques but is employed differently. The dog is cued to conduct an obedience task and rewarded for the correct behavior with a bite, slip, and carry. Use of this training method will create positive focus on the handler and higher drive in obedience. The end result provides a more reliable and confident animal. This method also helps to eliminate hesitation problems.

Continuation training

Employ the reward bite method intermittently throughout training. Use it to build the dog in all aspects of patrol training, including building search and scouting. When you use the reward bite in these scenarios, the decoy should work the dog to get a full hard bite and then release the sleeve. The training supervisor predetermines the level of "fight" the decoy uses, always striving to build the dog's confidence and reliability.

False Run (Decoy)

The objective of this exercise is to condition the dog to remain in position and not bite unless commanded to. Conduct initial training on leash. When the handler commands the MWD to stay, begin to suspiciously advance toward the dog. Use successive approximation and move toward the dog from the starting position. The dog's actions will dictate how many trials it will take to completely train the dog. If the dog attempts to bite, or fails to remain in position, the handler must immediately correct the dog. To prevent the dog from becoming deficient in aggression and attack, the agitator and handler should decide when to give the dog a bite. To maintain aggressiveness, you may allow the dog to bite on a random basis during this type of training. To fully condition the dog, the decoy should mimic provocative behavior encountered in real-world situations.

Controlled Aggression

Controlled aggression is used to teach the dog to pursue, bite, and hold on command. The team starts in the HEEL/SIT position off leash. Wearing the arm protector, move around suspiciously about 40 to 50 feet in front of the team. The handler will order the agitator to halt and place hands over head. The agitator ignores this order, turns, and attempts to run away. The handler commands the dog "GET 'EM." When the handler calls the dog "OUT!" the decoy ceases all resistance and agitation.

Standoff

This training enables the handler to gain complete control over the dog after commanded to pursue. The starting position is the same as with the attack and apprehension. Approach the dog making provocative gestures. When you get within a few feet, turn and run away. After you're about 30 feet from the team, the handler will command "GET 'EM." When you hear this command, cease all movement. The dog will be called "OUT." This training may become confusing to the dog; therefore, to keep it at an acceptable level of aggressiveness, allow it to bite at irregular intervals. **NOTE:** You may vary time and distance in all aspects of standoff training depending on the dog's proficiency level.

Double Decoy Attack

This exercise requires an additional decoy. The purpose of this exercise is to teach the dog to ignore one of the decoys while pursuing, biting, and holding the other. The dog starts off leash in the HEEL/SIT position. Position the decoys approximately 30 feet from the dog team. The handler challenges by ordering them to halt. One decoy obeys the command while the other ignores it and runs away. The handler immediately commands "GET 'EM." The dog ignores the decoy that halts and pursues, bites and holds the second decoy. During the early stages of this training, attract the dog's attention by making provoking gestures and noises.

Scouting

The primary mission of the MWD is to detect and warn the handler of the presence of an intruder. The team is placed in a semi-cleared area facing into the wind. The terrain features in front of the team should allow the decoy to run and crouch behind bushes and trees. Before the decoy starts to run, the handler tells the dog to "WATCH 'EM," the decoy will run from one point to another, acting suspicious, and hide at a pre-designated position of cover. The decoy will leave cover and run when the team is within 15 feet. This exercise is concluded with a short chase and bite.

Building Search

On-leash building search

Introduce the building search on leash as an agitation exercise ends in such a way that it will seem a natural extension of agitation training. For example, as an agitation exercise ends, the agitator runs away from the dog and hides behind the doorway to an adjacent building. The team will pursue to the doorway. The agitator continues to provoke the dog to illicit a desire to pursue. The procedures remain the same until the dog is ready to advance to the next step, which is to enter a building and actually seek out an intruder while concealed on floor level. To assure the dog of the agitator's presence, you may need to provide a faint noise or an obvious movement for the dog. The noise and/or actions should cause the dog to bark or produce a response. The decoy may have to agitate or act afraid of the dog to get a favorable response. Conclude this exercise with the team escorting the agitator out of the building.

Off-leash building search

During off-leash building search, perform exercises in the same manner as on leash. Conceal the agitator in a location inaccessible to the dog—either floor level or an elevated position. The agitator must remain quiet and motionless allowing enough time for the dog to detect and respond. The agitator may need to make noise or partially reveal himself to ensure success. The ultimate goal is for the dog to detect and respond to the agitator while separated from the handler.

PROFICIENCY STANDARDS AND EVALUATIONS

The kennel master is responsible for establishing an effective training and evaluation program to maximize the dog and handler's proficiency. The post-certification standards establish minimum proficiency standards the dog team must maintain. These standards must be met within 90 days of team assignment and validated annually thereafter. Certification standards are a combination of dog training scenarios conducted within the controlled environment of the MWD section's dog training area and within the actual working environment the dog team performs their duties. Certifications are documented on the AF Form 321/Military Working Dog Training and Utilization Record. Should a team fail to meet the minimum proficiency standards, the kennel master will immediately initiate remedial training. If, after remedial training, it is determined the poor performance of the team is due to the dog and not its handler, contact the 341 TRS Dog Training Section for further guidance.

Obedience Commands
Dogs must respond to the handler's commands of "SIT," "DOWN," "HEEL," and "STAY" at a distance of 50 feet with no more than one correction per five commands. The dog must, on the command of "STAY," remain either in the SIT or DOWN position for 3 minutes.

Controlled Aggression
Evaluations should be a combination of scenarios conducted within the controlled environment of the MWD training area and on post.

False run (off leash)
Given the command "STAY," when confronted by an individual who approaches no closer than 3 feet, the dog must not break position.

Standoff

When commanded to bite and hold the intruder, the dog must pursue until commanded "OUT" and "HEEL" by the handler. The dog must not bite the intruder regardless of the intruder's actions.

Bite and hold

When commanded "GET 'EM," the dog must pursue, bite, and hold an intruder for a minimum of 15 seconds. The dog must not release until commanded "OUT." Dogs must demonstrate proficiency in this task with either an exposed or concealed arm protector.

Search and escort

The handler positions the suspect at a distance of 6 to 8 feet from the dog, facing away from the dog and advises the suspect not to move. The handler then moves up and conducts a search of the suspect. Once the search is conducted, the handler positions himself directly behind the suspect(s) and calls the dog to the HEEL position. The handler then puts the dog on leash, takes control of the suspect by placing his non-leash hand on the suspect's shoulder and then conducts the escort. Prior to initiating the escort, the handler will instruct the suspect to only move when told to do so, and that any sudden movement of the suspect may result in the dog breaking the HEEL position and biting. The dog may heel on the handler or slightly forward and to the side of the suspect to ensure an effective escort. During training, re-attack scenarios should be conducted periodically to ensure that the dog will perform the task without command.

Building search

The dog finds an intruder with or without an arm protector hidden in a building and indicates to the handler the presence/location of the intruder.

Scouting

The dog finds an intruder hidden in an open area or field by scent at 50 yards, sight at 35 yards downwind, or sound at 35 yards downwind. Consider terrain and weather conditions when evaluating by these standards.

Vehicle patrol

The MWD rides in a vehicle driven by the handler without showing aggression toward the handler or other passengers.

Gunfire

The MWD successfully performs basic obedience, and controlled aggression tasks, during gunfire. Gunfire during aggression phases of training must be kept to a minimum.

Obedience course

The dog negotiates the obstacle course at a moderate rate of speed on or off leash in the HEEL position.

SF STANDARDIZATION AND EVALUATIONS

The kennel master should work closely with the unit's STAN-EVAL section to assist in the coordination of practical/performance evaluations of dog handlers being formally evaluated as an MWD patrolman.

CHAPTER 4

CLEAR SIGNALS TRAINING METHOD

INTRODUCTION

Dog training methodology is not static. Instead, as a result of new insights and new information, dog training methods evolve, becoming more effective and more powerful. This chapter is meant to provide a brief review of new techniques that are available for training MWDs for obedience and controlled aggression. These techniques are currently used at the 341 TRS, Lackland AFB, TX.

CLEAR SIGNALS TRAINING METHOD

In the last 10 years, enormous technical progress has been made in the methods used to train working dogs for obedience and controlled aggression. The most important advances were made by amateur trainers who compete in obedience competition. The method developed for use by the DoD MWD Program is called Clear Signals Training (CST).

Clear Signals Training

CST is founded on three very important ideas: a) Teach skills with rewards, not physical force; b) Establish clear communication; and c) Use compulsion only when necessary, and use it in a fair and effective fashion.

Teach skills with rewards, not physical force

As much as possible, MWDs should be taught and motivated to work using rewards to induce desired behavior, rather than using force to compel the dog to do as the trainer wants (see Chapter 2 for discussions of inducive versus compulsive training methods).

Establish clear communication

One of the most critical aspects of dog training is the development of clear communication between handler and dog, so that the dog knows what the trainer wants, and it fully understands the relationships between its behavior and various consequences.

- CST makes sure the dog understands what the trainer wants by breaking training into stages, the first of which is a *teaching phase* in which the dog performs for rewards in a very low-stress atmosphere.

- CST makes sure that the dog understands the relationships between its behavior and specific consequences (see "Response Contingency" in Chapter 2) by using conditioned signals or "markers" (the words "Yes," "Good," and "No") to help the dog realize exactly which of its behaviors resulted in a particular reward or a particular punishment. This approach has only recently become established in the working dog world, but for many years it has been extremely influential in the training of exotic animals like killer whales and big cats. In more recent years the response marking approach has also influenced trainers that work with dogs assisting deaf persons and otherwise disadvantaged populations. When combined with the use of positive reinforcement techniques, response-marking methods are the most powerful available technology for shaping behaviors, and they should become a basic part of the DoD MWD Program skill set.

Compulsion use

Use compulsion only when necessary, and use it in a fair and effective fashion. CST assumes the dog trainer will use physical force or compulsion (see Chapter 2) in a way that is effective and fair, and the dog must first pass through two stages of learning. It must first be taught the skill (e.g., sitting or lying down), so that it understands what is required before it is subjected to any physical or psychological pressure. Then the dog must also go through another stage in which it is taught about the correction that will be used to apply the pressure (e.g., the jerk on the choke collar). In this stage the dog learns that it can terminate the correction by using a certain behavior, and it learns not to fear the correction (here we are discussing a very gentle version of escape learning). Only when the dog understands the skill and understands the correction and how to respond to it is it fair and effective for the handler to apply strong pressure (to generate avoidance learning).

Teach, Train, and Proof

CST breaks MWD training into a three-stage process. Each skill is first taught, then trained, and then proofed. Not all skills that we teach working dogs participate in all three of these phases, but many of them do, especially obedience skills like sit, down, and heel.

Teach

The initial stage is called teaching. During this stage the dog learns what is expected of it in a given situation. As a rule, teaching proceeds best when the dog works for a reward like food or a ball/Kong, and is not stressed or anxious. Earlier we described this kind of training as "inducive" and pointed out that the two inducive tools available are positive reinforcement and negative punishment (also called omission). During teaching we concentrate on motivating and teaching the dog in a low-stress atmosphere, giving it rewards when it performs correctly (positive reinforcement) and withholding rewards (negative punishment) when it makes mistakes.

- For example, we teach the dog to sit by holding food over its head. In the effort to reach the food, the dog lifts its head and rocks backward, accidentally assuming the sit position, and then we reward this action by allowing the dog to have the food.

- For example, we teach the dog to maintain eye contact with the trainer by making a noise so that the dog looks at the trainer's face, and then we allow the dog to have the ball/Kong.

- During teaching we avoid the use of any physical force or input designed to compel the dog to perform a specific behavior. Instead the dog is given the freedom to experiment with its own behavior and learn what responses bring reward and which responses do not bring reward. Errors are seen as desirable because it is by making mistakes that the dog sharpens its understanding of "correct" behavior.

- During initial *teaching* we normally rely on continuous reinforcement in which we reward every correct repetition of the exercise.

- Teaching normally involves the use of obvious gestures and body language cues that the trainer uses to lure the dog into position with the reward—moving the hand upward for the sit, bending at the waist and placing the hand on the ground for the down, etc. These gestures are called "prompts," and with further training they are normally faded out.

Train

The next phase of CST is called training. During the training phase we take the practiced skill that the dog has learned to use to obtain reward and we put it under the influence of some form of physical "correction." In dog training, it is very common to use the term "correction" for physical inputs that are meant to pressure the dog into certain actions. Thus "correction" refers to the other two response contingencies described earlier, negative reinforcement and positive punishment. This does not mean that during training we use heavy psychological and physical pressure on the dog. Far to the contrary, during training our main objective is to keep the dog comfortable and positively motivated, while showing it the "meaning" of each specific correction, teaching the animal that it has the power to control this correction, and teaching it how we wish it to behave under the correction.

- For example, we train the dog to associate a tug on the leash/choke collar with sitting by asking the dog to sit in the presence of food, much like before, but after we give the "Sit" command we give a soft correction on the collar by "popping" the leash. The pop does not make the dog execute the sit; the animal is already sitting because of habit and its desire for reward. The pop does not hurt the dog. The procedure "connects" the correction to the sit; it puts the sit skill under the control of the collar correction.

- For example, we train the dog to maintain eye contact (attention) by bringing the dog "into focus" on the handler, and keeping it there for a moment. An assistant provides a soft distraction like tapping a foot so that the dog looks away from the handler for an instant. At that moment the handler says the dog's name and an instant later gives a soft pop on the leash/collar. The pop does not make the dog look back at the handler. The dog looks back simply because it hears its name and it wants reward. But the procedure connects the "pop" to attention so that the dog's attention comes under the control of the leash correction.

Proof

The last phase of CST training is called *proofing*. During proofing the handler reduces the frequency of reinforcement (moving into intermittent and random reinforcement schedules), begins to fade out gestures (prompts) that have been used to assist the dog to execute skills, and begins asking the dog to perform the skills in different, distracting environments. As a result of these changes, the dog's performance is disrupted. To put it another way, the animal makes mistakes, and these mistakes may be corrected (to generate avoidance learning).

Corrections administered during proofing of a particular skill, like sitting under distraction or holding the down-stay, must be "fair." Another way to say that a correction is "fair" is to say that it is effective and the dog learns very quickly to avoid any future corrections. Providing that we have done a good job of teaching and training a skill, proofing of that skill normally proceeds efficiently (i.e., results in avoidance responding) and without any particular upset or stress on the dog's part.

Table 6. Teaching, Training, and Proofing Phases of Various MWD Skills

	Teaching Phase	Training Phase	Proofing Phase
Skill	Dog learns to perform a skill for reward—errors are permitted and even encouraged so that the dog learns which behaviors are successful and which behaviors are errors.	Dog learns that the skill terminates or "turns off" a gentle correction, and that an error (such as breaking attention or breaking the sit-stay) "turns back on" the correction See *escape learning*.	Dog learns that, even though the skill still earns reward, this reward is less frequent, and the skill is mandatory—Refusal and errors "turn on" sharp corrections, and compliance to first command avoids corrections altogether. See *avoidance learning*.
Attention	Teach dog to look at the handler on cue (usually the dog's name) for food, then transition to ball/Kong reward.	Train dog to "turn off" mild collar corrections by looking at handler on cue, or by continuing to look at the handler.	On first command, dog must pay attention to the handler, and maintain this attention despite distractions, to avoid sharp collar correction. Initially dog is given food or a ball/Kong frequently to reduce stress; later these primary rewards to some extent are "weaned out."

Down	Teach dog how to lie down for food, then transition to ball/Kong reward.	Train dog to "turn off" mild collar correction and "social correction" (see text below) by lying down, or by holding the down.	On first command, dog must lie down quickly and hold the down, in order to avoid a sharp collar correction or strong social correction (see text below). Initially dog is given food or a ball/Kong frequently to reduce stress; later these primary rewards to some extent are "weaned out."
Heel	Teach dog to move to heel position for food, then transition to ball/Kong reward.	Train dog to "turn off" mild collar corrections by moving to heel position, or by staying at heel and in attention.	On first command, dog must move to heel position, establish attention (looking at handler's face), and maintain heel position and attention to avoid sharp collar correction. Initially dog is given food or a ball/Kong frequently to reduce stress; later these primary rewards to some extent are "weaned out."

| Out | Teach dog to release grip on "dead" object to earn another bite and "fight" — not always practical.

Normally begun using a less motivating object such as PVC pipe. | Train dog to "turn off" collar corrections by releasing bite. If necessary, dog forced to release with collar correction.

Performed with more motivating bite object like rubber hose or jute tug toy. | Dog must release bite immediately on first command to avoid sharp collar correction.

Performed on decoy with bite sleeve, suit, etc. |
|---|---|---|---|
| Standoff | Teach dog to earn bite by lying down on command. —decoy comes to the dog to give reward. | Train dog to "turn off" mild collar corrections by stopping forward movement, and then lying down. | Dog must stop forward movement on command to avoid sharp correction—down maintained by positive reinforcement—i.e., once dog stops moving forward, tends to lie down voluntarily to earn bite. |
| Odor Recognition | Teach dog that target odor is associated with ball/Kong. | | |
| Final Response | Teach dog to sit to gain access to a blocked ball/Kong ("blocked" = held in hands, wedged between furniture and wall, etc.). | Sit in response to odor supported by the use of mild cues (pressure on rump, mild pops on collar) that are "turned off" by the sit. | |

"Escape Training" Method

We should compare CST to more traditional methods in which the dog is not taught skills inducively, but instead has to learn what is expected under physical force. In this situation, because the skills are associated from the beginning of training with psychological pressure and discomfort, the dog learns to dislike its work and to resist the trainer, which makes the animal difficult to train. More important, the dog also becomes defensive. What we mean by defensive is this: When we expose an animal time and again to something unpleasant like a jarring collar correction, or choking with a chain collar, and we have not previously taught the animal how to understand this correction, then the first thing that the animal begins to do when it anticipates a correction (e.g., when it hears the "Sit" command) is protect/defend itself by stiffening its neck and bracing its legs. Obviously a dog that is physically stiff and tense has great difficulty sitting quickly (or doing anything quickly), so the dog tends to make more mistakes, and receive more corrections, which makes the animal even more stiff and tense, and so on. For these reasons, traditional "escape" methods for obedience and patrol tend to generate very high levels of resistance, fear, and confusion on the dog's part.

In contrast, CST is directed at preventing fear and confusion by first teaching the dog exactly what to do on command to get a reward (teaching), then teaching it how to use this action to "turn off" a correction (training), and finally showing it that sometimes it has no choice but to do what its handler commands (proofing). When training is approached this way, even if the trainer finds it necessary to apply physical force to the dog, the animal is not excessively frightened or stressed because it knows what to do and how to control the correction it receives (respond quickly to terminate the correction, and then in the future to behave so as to avoid the correction completely).

The "Clear Signals" Part of CST

The division of MWD training into teaching, training, and proofing phases is a new idea in the DoD MWD program, but it is not cutting-edge theory. Teach-train-proof is a version of what the best dog trainers have been doing for decades now. But the next part of CST is more revolutionary—it involves the use of conditioned cues to communicate with the dog, to give it very precise information about the relationship between its behavior and rewards and punishments. These conditioned cues are called "markers," and often "bridges," because they perform two very important functions—marking responses for reward or correction, and bridging delays to reward or correction.

Response (or behavior) marking

Most skills in dog training feature a critical aspect, a point in the course of the skill when the trainer's requirement is fully met. This can be the moment the dog locks its eyes on its handler's in response to the "look" command for attention, or the moment the dog's elbows touch the ground in response to the "Down" command. If we can make the dog understand that it is these critical aspects, these core requirements of the skill, that earn reward, then we can become very effective trainers.

In traditional training we try to make the dog understand by following the old rule "in order for the dog to understand why it is being rewarded or corrected, reward and correction must follow immediately after the desired behavior." The problem is that in many situations it is very difficult to reward a desired behavior without accidentally rewarding some behavior that comes after the desired behavior. To choose just one of very many examples, if I am teaching my dog to march/heel while maintaining attention on my face, and I start to reward this behavior with the Kong in my right pocket, when the dog sees me move my hand toward my right pocket it looks at my hand instead of my face, and often it crosses in front of me to follow the hand to where it knows the reward is. Therefore, when I produce the ball and give it to the dog, I am not rewarding the behavior I want (clean heeling at the left side with eye contact) but other behaviors (looking at my right hand, forging ahead, crossing in front, and interfering with my movement), which are detrimental to my goal.

- *The "Yes" release marker.* The use of a marker solves this problem for us. The marker we most commonly use is the word "Yes" said in a distinctive way. To use the "Yes" we must first condition it—turn it into a secondary reinforcer by pairing it with the reward (see discussion in Chapter 2 on classical or Pavlovian conditioning and secondary reinforcers). Most often we use "Yes" with ball reward. We can condition the marker by saying "Yes" and then giving the dog the ball about a dozen times, with about ½ to 1 second between "Yes" and the ball. Once the "Yes" marker is conditioned, then it has gained the power to act as a reward. The word has become significant, and when I say "Yes" immediately after some behavior on the dog's part, the dog "notices" what it has done—the behavior is marked.

To produce rapid learning of the desired behavior, you need to let the dog have the primary reinforcer, the ball. And here is the advantage to using the "Yes": Because I have marked the desired behavior with the word "Yes," you are no longer under pressure to get the reward to the dog quickly, and you do not need to worry about any behaviors the dog engages in between the behavior you want to reward and when the dog actually receives the ball. The "Yes" marker does three critical things for the trainer.

1. "Yes" marks a specific desired behavior (e.g., making eye contact while moving in heel position).
2. "Yes" releases the dog from the behavior—when the dog hears "Yes" it knows it's finished with its job.
3. "Yes" bridges the delay to reinforcement (hence the term bridge, often used interchangeably with marker). Even though it may take me 10 seconds to get the ball out of my pocket and give it to the dog, and the whole time the dog is dancing around me and jumping up and down in anticipation of getting its reward, when the dog actually gets the ball it will associate this reward with what it was doing right before it heard "Yes" instead of what it was doing right before it got the ball.

Because the dog can release after hearing "Yes," because the job is ended when the "Yes" is given, this cue is called a "terminal marker." Saying "Yes" is just like saying "OK," in that the dog can do anything it wants after it hears "Yes."

- The "Good" marker. However, sometimes we don't want to release the dog, we want to encourage it but keep it performing. In this case we need a different kind of marker. Because the skill is not done when we use this marker, it is called an intermediate marker (as opposed to the terminal marker, "Yes"). In CST the intermediate bridge is usually the word "Good," said in a distinctive and encouraging voice.

The sequence runs like this: While the dog is performing the skill it receives one or two "Good" markers at critical moments, then when it has finished the skill to the trainer's satisfaction it receives the "Yes," and then the primary reward. Because the dog hears "Good" before receiving the "Yes" cue and then receiving the reward, after a number of repetitions the "Good" also becomes associated with reward (this is called second-order conditioning) and "Good" also takes on the power to reinforce and mark behaviors.

We have now described two different ways of rewarding the dog with markers: "Yes"-release, and "Good"-marker followed by "Yes"-release.

- *"Yes"-release.* When the dog has completed the skill the trainer rewards it with the "Yes" cue. On hearing "Yes" the dog breaks from position, and the trainer provides the primary reward. Yes-marking is used to teach the dog to respond swiftly to commands.

- *"Good"-marker followed by "Yes"-release.* When the dog is doing well, the trainer encourages it with the "Good" cue, the dog continues to perform, and then when the skill is complete, the handler releases with "Yes," and then provides the primary reward. "Good"-marker then "Yes"-release is used to encourage the dog and keep it performing.

Reward in position

A special kind of "Good"-marker followed by "Yes"-release is used to stabilize the dog into certain positions, such as the sit and down at the end of the leash, sit at heel position, and the field interview. These skills are all characterized by one problem—the dog is supposed to stay in a particular place and position, but it will be rewarded from another location. For instance, after sit and down EOL, the dog is recalled to heel and then released and rewarded. Because the dog anticipates this reward, it begins to creep forward toward the handler while moving from down to sit or vice-versa, and many difficulties can follow. To prevent these difficulties with anticipation, the handler can follow this procedure: When the dog accomplishes a correct transition (e.g., rising from down to sit without moving toward the handler), the handler marks this behavior with "Good!" Then the handler walks to the dog, removes the primary reward from his pocket, and holds the reward very close to the dog's nose while the dog maintains position (obviously some preliminary training is required to gain good control of the dog's behavior in the presence of the reward). When the dog is steady the handler gives the "Yes" cue; the dog normally first takes the reward and then breaks from position. Because the dog receives the reward in position, the animal is not encouraged to creep while performing EOL. Because the handler used the "Good" to mark the correct transition from down to sit, the dog knows what it is being rewarded for. As a result, with repetition the dog will become more and more proficient at performing EOL without creeping toward the handler. This method of providing reward while the dog is in position, and using "Good" to make sure it knows what it is being rewarded for, is called *reward in position*.

"OK"

This is another method of releasing the dog from work. For clarity of communication, it is extremely important that the dog know when a particular skill is finished. The verbal cue that the trainer uses to tell the dog that a skill is finished is called the release. We have already discussed how the "Yes" marker serves as both a behavior marker and a release cue. The trainer can also release the dog with the cue "OK." The difference between "Yes" and "OK" is that while "Yes" is a promise to the dog that it will be rewarded after a delay, "OK" is used to release the dog from work when the trainer does not intend to give the dog a primary reward. This can be the case when the dog will merely be praised, or when the dog has made a mistake, and the trainer intends to make it repeat the exercise correctly before providing a reward.

Rewards Used in Obedience Training

CST (like practically every effective system for teaching obedience to working dogs) relies heavily on positive reinforcement to teach lessons and motivate performance. Sources of positive reinforcement are praise, food, tug toys, and ball or Kong.

Praise

Praise and social reinforcement are vital ingredients to working dog training, and some dogs can be obedience trained with nothing but praise as a positive reinforcer, combined with corrections to discourage disobedience. The problem is that few of the dogs that DoD procures are trainable with this method—they are not socialized in the way that pet dogs or sport dogs are socialized, and without a background of proper socialization, praise alone is often not sufficient to support efficient training. In addition, praise suffers from one great disadvantage. Praise does not provide us with a focal point or goal that we can use to attract, lure, and manipulate the dog, the way we can with, for instance, a handful of food.

Food

Food is a very effective positive reinforcer for many dogs, but eventually we must wean the dog off of food and find other sources of motivation that are more operationally practical. In addition, a substantial number of the dogs procured by DoD do not have enough desire for treat food to perform useful training.

Tug toy

Many dogs work well for the opportunity to play with a tug toy, but not all. Dogs that are bought for detection only may exhibit little desire to bite and tug on an object.

Ball

This leaves one remaining source of motivation–ball or Kong (hereafter referred to as "ball"). This is just as well, because the ball is cheap, easily carried and used, and every military dog is selected especially for its intense desire to chase, carry, and play with balls. Many DoD trainers (educated in the "escape" method of obedience training) never allow an MWD to play with its ball except in the course of detection training, for fear that ball-play will devalue the dog's primary reward for detection training. However, as long as the dog has high levels of retrieve/play drive, this is an outmoded and unnecessary practice. Countless highly effective detector dogs are allowed to play with the ball in obedience and sometimes just for exercise and the sheer fun of it, and they do not lose their effectiveness in detection. On the contrary, ball play keeps their enthusiasm for the ball high, and their physical condition and stamina high as well.

Corrections Used in Patrol Training

CST resembles traditional methods of dog training in that it makes extensive use of physical inputs called "corrections." Corrections are designed to do three things for the trainer: a) Increase the dog's precision of performance, b) reduce dependency on primary rewards (food, ball/Kong, and tug toy), and c) ensure that the dog performs correctly regardless of distractions. Fundamentally, corrections accomplish these functions by exerting punishment and negative reinforcement effects, which means that by definition effective corrections are unpleasant for the dog. The responsibility of the trainer is to use this unpleasantness for proper effect, while treating the animal fairly and humanely, and ensuring that, although moments of the dog's training are unpleasant, on the whole it enjoys training, and has affection and trust for the trainer rather than fear.

Humane versus inhumane corrections

Barring techniques that are likely to produce physical injury, it is not possible to categorize certain forms of correction as humane and others as inhumane simply on the basis of the physical parameters of the corrections. The primary concerns in judging whether a given correction technique is humane in a certain situation are:

1. Will the correction cause physical injury? A correction procedure that causes physical injury to the dog, or is likely to cause physical injury, is inhumane.
2. Does the dog understand what behavior is required of it in this situation? If the dog does not understand what behavior is required, or cannot very quickly learn the required behavior, then the procedure is inhumane.
3. Is the dog capable of executing the desired response in this situation? The dog may be incapable of executing the response, even though it understands the response (i.e., it may not be able to sit quickly, because it is too physically tense and apprehensive to do so, or it may not be able to release a bite object on command because severe treatment has conditioned biting to the pain of collar corrections). If the dog is incapable of executing the desired response, then the procedure is inhumane.
4. Does the dog learn from the correction, so that it quickly changes its behavior and thereby avoids further corrections? If the procedure does produce learning, with the result that the dog

continues to experience correction in training session after training session, then it is inhumane.

Defining humane and inhumane corrections

By the above definitions, procedures involving rather mild corrections may be inhumane, because they are not effective for one reason or another and the dog never learns to prevent them. Even mild events, if they are chronic and unpleasant and uncontrollable, can cause significant suffering. By the same token, procedures that appear rigorous and severe may be eminently humane because they do not injure the dog, and because they result in rapid learning and no more corrections. These are subjective judgments best made by experts in dog training, but these experts must always be prepared to justify their procedures and practices on the basis of the numbered list above. Ultimately, perhaps the best indicator of what is humane and what is not is the dog. Is the animal eager for work and eager for contact with its trainer at all times? Then it is likely that this dog's training is conducted humanely. Or does it consistently show inhibition, avoidance behavior, and fear in training contexts? Then it is likely that at some point the dog was treated inhumanely.

Tools and methods for correction

1. *Choke collars.* Choke collars may be made of nylon webbing, light or heavy chain, or nylon cord, connecting two metal rings. In general, the thinner and smaller the chain or cord, and the more efficiently it runs through the rings, the more severe a collar it is, because forces exerted through the collar on the dog's neck are distributed over a smaller area. The choke collar is used either to make a jerking correction followed by an immediate release of pressure (i.e., a "pop"), or by exerting a steady pull to produce a choking sensation. Choke chains are very well accepted in American life, and are sold wherever pet supplies are sold.

2. *Pinch (or prong) collars.* Pinch/prong collars are assemblies of heavy, bent wire links arranged with a chain yoke, so that when the leash attached to the yoke is pulled, the links tighten on the dog's neck and the dull ends of the wire links exert pressure on the dog's neck. They produce a sharper sensation than

choke collars do and can be used with a lighter touch and more precise timing, because less force is needed to get an equivalent effect. Pinch collars are generally used by jerking or popping the attached leash. They are also very well accepted in American life, and are sold wherever pet supplies are sold.

3. *"Social corrections."* Social corrections are slaps or cuffs of the foot, hand, or leash-end, or pokes of the fingers. In many circumstances, the quickest and most efficient correction is made by gently slapping the dog with hand, foot, or leash-end, or by poking the dog with stiffened fingers, especially when the handler's intent is to stop the dog from moving forward or biting. Such corrections have the advantage that they do not depend upon the presence of leash and collar, and therefore if used properly give the handler greater control over the dog in a wider range of situations. Appropriate, effective, and humane examples are:

- The handler commands the dog to lie down while the MWD team is heeling rapidly forward (as when an MWD team is running from one position of cover to another in a fire zone). The dog does not lie down quickly, and the handler slaps the dog on the back or neck or ears with the hand or the leash end. The handler then heels forward again rapidly and repeats the command and, if the dog complies rapidly, it is praised and petted and given food reward while in the down position.

- The handler commands the dog to release a ball, the dog does so, and the ball drops to the ground and comes to rest. The handler commands the dog to "stay" and reaches to take the ball, but as he does so, the dog attempts to bite the ball, and thereby the handler's fingers. The handler says "No" and cuffs the dog sharply on the side of the muzzle with the open palm of the other hand. As a result, a moment later the dog is somewhat tentative in taking the ball from the handler's hands, even when invited to do so, and the handler praises and encourages this respect for his hands/fingers.

- The handler holds the dog on a 6-foot leash and gives the "heel" command, but the dog is distracted by nearby activity and reluctant to obey, and keeps turning its head and forequarters away and pulling into the leash, making it difficult for the handler to get enough slack on the leash to give a "popping" leash correction; so the handler uses the instep of his foot to slap the dog sharply on the big muscle at the back of the thigh. The dog, startled, turns to look at the handler, and the handler instantly encourages the dog to come by praising it and running backward, and then gives the dog a ball reward.

4. *Electronic/electric collars.* Electronic collars are devices that deliver a high-voltage but very, very low-amperage electric shock to the skin of the dog's neck through electrodes. They are operated remotely by means of a hand-held transmitter on which there are intensity settings and buttons that trigger continuous or momentary shocks through the collar and sometimes beeps and tones (markers) that aid the dog in understanding what is required of it. This electrical shock is like a static electrical discharge from the carpet; it is similar to, but much weaker than, the shock delivered by a livestock fence. This shock may be uncomfortable, but it does not cause injury, because injury is a result of the amperage of an electrical shock, not the voltage.

- Electronic collars are common and well-accepted instruments, not only among the pet-owning American public but also by Federal agencies (e.g., United States Secret Service) and very many state, county, and city law enforcement agencies.

- Electronic collars provide the ability to deliver electrical stimulation that is finely calibrated to the individual dog's level of sensitivity, over long distances and with very precise timing. However, electronic collars demand a very high level of technical knowledge and ability from the trainer, for a number of reasons:

  - Because the onset and offset of electrical stimulation is so precise and "clean," the electronic collar magnifies the effect of any errors of timing on the trainer's part. A satisfactory analogy is "razor blade"—it cuts clean but it had better be handled with expertise and care, or accidents will happen.

  - Without prior *teaching* and *training* to show the dog how to respond to it, electrical stimulation of the dog's neck produces something like a startle response—the dog throws its head up, or bends its neck and perhaps jumps forward or up, precisely as you would do if you were momentarily shocked on the back or neck by a prank buzzer. This means that the dog's natural reaction to electrical stimulation of its neck does not help it do any of the things that we might be interested in training an MWD to do—recall to the handler, lie down, or release a bite. Therefore, the dog must be very expertly prepared through the teaching phase (teach it what to do and how to do it to earn reward) and training phase (show it how to use this behavior to turn off very mild levels of electrical stimulation) of CST, before the proofing phase (convince it that it must perform the behavior to avoid uncomfortable levels of electrical stimulation) is accomplished with an electronic collar. Otherwise even the most well-meaning trainer can completely fail the "humaneness" test. The dog may not understand what it is supposed to do, be unable to do what it is supposed to, and be unable to learn what it is supposed to do. With even mild levels of shock, this is inhumane and unacceptable.

- As of this writing, the only arm of the United States Air Force that is authorized to employ the electronic collar is 341 TRS, Lackland AFB, Texas. The electronic collar is utilized only in particular cases with special authorization, under the supervision of specially trained personnel, to solve severe training problems in high-value MWDs.

OBEDIENCE TRAINING WITH CST

Sit and Down

Teaching sit and down
- *Sit* and *down* are best taught by luring the dog into position with soft, appetizing food. This food must be something like small pieces of meat or specialty dog food that the dog is eager for and that it swallows quickly without chewing. Crunchy treats do not work well. The dog is first taught to eat from the hand and then taught to maintain soft contact with the hand and follow the hand until it is allowed to eat. Then the handler uses the closed hand to lure the dog into position (square, erect sit; or sphinx-like down position with both elbows in contact with the ground), and then loosens the hand so that the dog can lick and nibble the food out of it while holding position. Then, before the dog breaks position the handler releases the dog by saying "OK" and enticing the dog out of position. Through a number of steps, this is developed into sit and down on command with stay (for 5 or 6 seconds), using the dog's food motivation only.

- The same method can be used with the ball, but because the dog's level of excitement will be much higher than in the case of food, the technique requires more skill and experience. In addition, prior to luring the dog into position using the ball, the animal must be taught to release it cleanly on command ("out"), to refrain from biting the reward until given permission, and to respect (i.e., not bite) the trainer's hands. The handler must be able to hold the ball/Kong in his hand an inch or two from the dog's head without having the dog take the ball until it is given permission. In addition, the dog must learn to follow the hand closely, the way a dog would naturally follow a hand in which food is held, but without snapping at or biting the hand.

- "Sit" and "Down" commands are given as the dog is lured into position. "Good" is used when the dog is holding position well, before the dog is fed. The dog is fed in position, without being released. "OK" and enticement are used to release the dog from position. "No" is used to mark errors, and to tell the dog that it will not be rewarded. For instance, the dog is in down position, the trainer moves the hand toward its nose to feed it, the dog begins to crawl toward the hand to eat, the trainer says "No" and stands upright, withdrawing the hand and the food until the dog restabilizes in down position. "Stay" may be used to steady the dog in position, once longer sits and downs are introduced.

- The "Yes"-release marker is not employed until the dog is proficient at sit and down, stays in position until it hears the "OK" cue, and has had many rewards in position. "Yes" is introduced by having the dog sit or lie down, saying "Yes," enticing the dog out of position (so that it releases), and then feeding the dog. "Yes" may be used with food reward or ball reward interchangeably.

Training sit and down

- The first training of sit and down begins with the stay component rather than the actual sitting or downing motion—that is to say, when we begin to prepare the dog for the experience of being forced to sit or lie down, we apply the force to make the dog stay in sit or down position, rather than sit or lie down in the first place. The handler uses a handful of food to lure the dog into position and then rewards the animal. Then he tells the dog to "Stay" and waits for a mistake. In fact, the handler does whatever is necessary to cause the dog to break the stay—stands upright, holds the stay an unusually long time, etc. When the dog attempts to break (prematurely release from) the stay, the handler applies a very quick but rather gentle pop on the leash (up and away from the handler in the case of the sit, and directly backward along the dog's spine in the case of the down), sufficient to stop the dog from breaking, and then the handler quickly brings the food hand back and feeds the dog and repeats the exercise. When this is done skillfully, it is not clear what keeps the dog in position, the collar correction or the dog's fixation on the food-bearing hand. With time the dog comes to associate the leash correction with sit and down position. Now we can begin to use collar corrections to enforce the "Sit" and "Down" motions, in addition to the stay.

- Alternate mode of correction for the down. For the down, especially, it is very advantageous for the handler to use a "social" correction (see above)—a slapping correction rather than a collar correction—because a slap is normally faster and if done well a slap is more effective in pressing the dog into the down position. But we cannot just suddenly slap a dog with leash or hand and expect it to understand. The animal must learn the meaning of this correction and connect it to the previously understood skill. The handler stands with food in hand and signals the dog into the down with a long hand movement toward the ground and past the dog's nose, with food in the hand. This is merely an exaggeration of the movement the handler normally uses to lure/signal the dog into down position with food. Once the dog is down, it is fed in position, then given the "Sit" command, enticed up to the sitting position, and then signaled back into the down. This sequence of sit-down-sit-down is repeated several times. After this repetition, the dog will anticipate the next "Down" command and it will be waiting eagerly to lie down. As the handler again makes the long downward hand gesture combined with the command "Down," he clips or cuffs the dog rather gently on the muzzle with the ends of the fingers. The dog will notice the contact, perhaps blink or flinch away, and then quickly lie down because both force of habit and the near proximity of the food will guide it into this well-rehearsed behavior. We must be clear—we are not forcing the dog to lie down. The animal is lying down voluntarily to obtain the food. Before it even feels the contact of fingers on its muzzle it is already beginning the down motion. But we are preparing the dog (training phase) for the experience of being forced to lie down (proofing phase).

Proofing sit and down

- During proofing of the sit and down skills, we begin to challenge the dog's understanding, with more distracting surroundings and longer stays, less frequent food or ball reinforcement, and more praise reinforcement instead. When the dog performs correctly it is rewarded and encouraged, and when it refuses commands or becomes distracted, the corrections that were introduced and "attached" to the exercises during training are used in a stronger form, to ensure compliance. For instance, if the dog refuses to sit it is given a quick, popping collar correction upward. If it refuses to lie down, the handler slaps it lightly but sharply on top of the neck or skull with the flattened hand or with a loop of the leash.

- As a rule, early in the proofing process, to keep the animal motivated and reduce stress, we tend to give the dog rewards after a correction. For instance, if the trainer gives a "Sit" command, but the dog is distracted by another dog nearby and therefore does not sit, the handler delivers a quick, popping correction on the leash/collar, the dog sits, and then the handler rewards (either with "Yes"-release, or "Good"-marker then-"Yes"-release [see "The 'Clear Signals' Part of CST" section earlier in this chapter]). Rewards after corrections help to reduce stress, and help the dog "keep trying" even under a little bit of pressure.

- Later in proofing we raise our standard. If the dog must be corrected to secure compliance, then the animal does not receive any reward beyond a bit of praise and petting. Then the dog is released and immediately asked to repeat the exercise. If this repetition is correct, then the dog is rewarded.

- We should always keep in mind the difference between a mistake on the dog's part, where the animal is trying to do as the trainer asks but just makes an error, and disobedience or refusal. As a rule we do not often correct simple errors, instead we punish them with the word "No" and we withhold reward (omission). Sharp corrections (positive punishment and negative reinforcement) are normally reserved for disobedience or refusal.

- During proofing of sit and down, the motivation is normally supplied by a ball. The dog is initially taught using food if possible and then once it understands the exercises, the ball is introduced. Introduction of the ball will result in the dog becoming much more excited than it did when working for food. Ball motivation is thus a good way to challenge the dog so that it makes a few mistakes, and also it helps the dog to shake off any discouragement or stress it feels as psychological pressure gradually becomes a part of training.

Communication during sit and down

- While initially *teaching* the sit and down with food, verbal cues are of relatively little importance—the handler's gestures as he lures with food are most important. However, we normally give the "Sit" or "Down" command as we lure the dog into position, praise the dog with "Good" before and during reward, and release with the cue "OK." Later, as we move into *training* and *proofing* stages and begin to use ball reward, these verbal cues become very important, and we also begin to make use of the markers "Yes," and "No."

- Reward in position for sit and down. Our first concern with sit and down is establishing stability—making the dog understand that its job is to stay still without fidgeting or creeping. The best way to do this is to make sure that the animal receives its reward while it is still holding the sit or down. This is called a "reward in position" (see above in this chapter). For both sit and down, reward in position is performed as follows: The handler, with the ball in pocket, gives the "Sit" or "Down" command. When the dog moves swiftly and correctly into the appropriate position, then the handler marks this behavior with the "Good" cue (intermediate marker). This cue tells the dog that it performed well and earned reward, but that it must not break position yet. Then the handler gets the ball out of his pocket and holds the ball very closely in front of the dog's nose. Some care and a bit of training is required so that the dog does not creep or break position as the ball is brought out, and does not try to take the ball before given permission. After a moment in which the handler makes sure the dog is steady, he gives the "Yes" cue (terminal bridge), which is the dog's authorization to take the ball. If this technique is done well, the dog takes the ball while still in position and then releases from the position. Use of "Good" and "Yes" cues, combined with reward in position, achieves the twin goals of making sure we mark and reward the correct movement into sit or down position, yet also keep the position stable and prevent creeping, fidgeting, or breaking toward the handler and reward. *Reward in position (using "Good" then "Yes"-release) teaches the dog to hold a position.*

- "Yes"-release for sit and down. Once we have a stable sit or down, our next concern is making sure that the dog understands that the correct response to the commands "Sit" or "Down" is a rapid, crisp movement. To do this we must reward the sit or down movement, rather than the stay. What is important here is that we pick out and mark the critical aspect of the skill—in the case of the sit, the moment the dog fully sits; and in the case of the down, the moment that the dog gets both elbows in contact with the ground. Using the down as our example, the handler does this by giving the "Yes" cue the instant both elbows are in contact. When it hears "Yes," the dog will release from position and wait to be rewarded. The handler then breaks position and retrieves the ball from his pocket and gives it to the dog. The handler must not be in a hurry to get the ball out, and it is extremely important that the handler not break his position until after he has said "Yes." The critical aspect of timing here is when the handler says "Yes." The "Yes" must be timely, but the delivery of the ball can be and should be done deliberately and without hurry. Remember that the function and value of the "Yes" terminal marker is that it bridges delays to reward. Therefore, if the "Yes" is properly conditioned and well timed, there is no need to hurry in delivering the ball. *Reward at completion of a movement (with "Yes"-release) teaches the dog speed in assuming a position.*

- "No" marker. "Good" and "Yes" are not the only markers we can use, nor is the power of the marking technique limited to rewards. We can also mark behaviors we want to punish, by using the word "No." "No" is used much like "Yes," in the sense that it is used to mark a behavior, and bridge a delay—in this case a delay to punishment. Earlier we covered two kinds of punishing response contingencies, positive punishment (giving the dog something that it dislikes) and negative punishment (or omission, taking away from the dog something that it likes). "No" can serve to signal both of these response contingencies. "No" is also used like "Yes" in the sense that it tends to be a terminal cue—when the dog does something that earns a "No," the animal often has to start the whole exercise again and therefore it can break when it hears "No."

- We can use the "No" to gently punish the dog (through omission) for over-eager or careless mistakes. For instance, if we are working the dog through a sit-down-sit sequence, and the dog does not wait for the sit command but pops up without permission, then the handler gives the "No" and makes the dog go back into the down again and wait for the command before rising up into sit. If the animal rises to the sit correctly, it can be rewarded in position with "Good" and "Yes," or it can be rewarded with "Yes" and allowed to break immediately. Here the "No" gives us the ability to improve performance without killing the enthusiasm of an eager dog.

- We can also use the "No" more forcefully as a predictor of physical (positive) punishment. Let us say the handler leaves the dog on a down-stay and steps a few feet away, and the dog breaks position and moves toward the handler without permission. If the handler then simply corrects the dog, the animal will be corrected in the act of approaching the handler, which can be hopelessly confusing for an eager dog. What is needed is a tool to tell the dog exactly what critical behavior "earned" it the punishment. Therefore, the handler gives the "No" immediately when the dog's elbows lift from the ground. Then the handler calmly approaches the dog and administers a correction of appropriate strength for the dog, normally by popping the leash two or three times, then takes the dog back to the exact place where the dog was lying and commands it to lie down. Then the handler steps away and, if the dog holds position correctly, the handler performs a reward in position—first "Good" to reinforce the act of holding position, then approach and placement of the ball directly in front of the dog's nose, and "Yes" to release the dog into the ball.

Advanced sit and down exercises
- To meet certification standards, the dog must eventually learn to transition from sit to down and back up to sit again at heel position, and while at EOL. For these exercises, the dog must not only understand sit and down, but also how to move from one position to the other without creeping forward or changing its alignment. This is where "Good" and "Yes" and "No" cues come into their own, because they give the handler the power to teach the dog to understand the difference between a perfectly correct down (in which the dog does not creep forward) and an incorrect down (in which the dog lies down every bit as fast, but creeps forward as it does so).

- For sit and down EOL, "Good" is used to let the dog know immediately when it has performed a correct sit or down, and then the handler approaches and delivers reward in position (by putting the ball close to the dog's head and releasing the dog into it with "Yes"). If the dog creeps as it transitions from sit to down or vice versa, then the handler marks the mistake with "No," and either makes the dog repeat the skill or calmly delivers a correction and then makes the dog repeat the skill.

- For sit and down at heel, "Good" is used in the same way to mark a correct transition. Reward in position is performed by taking the ball from the pocket and holding it directly in front of the dog's nose at heel position, and then releasing the dog into it with "Yes." If the dog creeps forward or slews sideways at heel while performing the transition, the handler marks the error with "No" (perhaps followed by a correction) and sends the dog back to correct heel (see "Teaching Heel" subsection "The Finish" below) in the original posture (sit or down). Then the handler makes the dog repeat the exercise, rewarding the dog in position if it is executed correctly.

- For real-world operations/utilization, the down has greater importance than the sit. The down is the dog's most stable position. When given the "Down" command, the dog must drop immediately no matter where it is and what speed it is traveling, and then lie still and silent until given another command or released, even under intense distraction (gunshots, decoys carrying bite equipment, cracking whips, etc.). Accordingly, substantial psychological pressure must often be applied to achieve this level of obedience. For such treatment to be fair and effective, corrections used for the down must be thoroughly trained (see "Train" subsection under "Clear Signals Training Method" above), and "Good," "Yes," and "No" must be properly applied so that the dog understands what it is being corrected for and can adjust its behavior to avoid corrections.

Heeling (Marching)

- Heeling is an attention-based exercise in which the dog walks at its handler's left side (personnel who carry weapons on the left side often teach the dog to heel on the right) with the shoulder even with the handler's knee, keeping pace and position no matter what the handler's pace or direction, and sitting automatically when the handler halts. The primary functions of heeling are to refine the dog's obedience to its handler and to provide the ability to transport the dog under close control through hazardous or distracting circumstances with both hands free. A well-trained dog concentrates completely on its handler while heeling, and heeling is therefore very fatiguing and not an appropriate way to transport the dog long distances. For mere transportation, where all that is important is that the dog remains under control, does not pull against the leash, and does not use any more energy than necessary, a "walk easy" skill is used instead. Walk easy (the command is normally "easy") is much less strict and demanding than heeling.

- It is traditional in DoD to use the verbal command "heel," and also slap the left hip with the left hand, and to repeat these verbal and gestural commands at each change of pace or direction. However, if we view heeling as a tactical tool rather than a parade-ground skill, then we must realize that: a) slapping the hip is unnecessary and inadvisable, because the left hand should be left free (for weapon-handling, for instance); and b) repeated commands are also unnecessary and inadvisable, because the verbal command "heel" is all that is necessary to tell the dog that it should place itself at heel position and remain there, no matter what the handler's movement or direction, until released. Accordingly, in the discussion of heeling that follows, the left-hip slap and repeated commands are not used. The CST method for teaching heeling described below normally results in a dog that positions itself for marching close to the handler's left knee and hip, with the handler's left hand hanging or swinging outside of the dog's head. In effect, the dog positions itself "between" the handler's left hand and hip and looks directly up at the handler's face, rather than positioning itself "outside of" the handler's hand and arm and looking around the handler's elbow at the handler. This close positioning is in most situations advantageous because it gives better and closer control of the dog and results in fewer training problems and less stress for the dog. Weapons retention is not normally an issue for law enforcement applications because the dog works on the handler's non-weapon-carrying side and the handler need not carry the left arm pinched close to the body (a left-handed handler normally trains the dog to heel on the right side and so can carry the right arm loosely while holding the left arm tight to the body for weapons retention).

- The CST method of teaching heeling concentrates on teaching the dog to understand the exact position that it must maintain at heel, how to move its body to reach that position, and on making sure that the dog is highly motivated for the work. This is the best way to prepare the dog for the physical corrections that may later be necessary to render heeling "fail safe" for real-world tactical scenarios, in which handler safety depends on his control of the dog.

Teaching Heeling

The finish

The dog first learns to heel not by walking at the handler's left side, but instead by learning to "finish." "Finish" is an expression used by competitive obedience trainers to describe a skill in which the dog moves from position in front of the handler to the heel position. In DoD, the dog does not normally walk around behind the handler, passing to the handler's right—instead the dog passes by the handler's left hand, turns in place, and sits at heel (called the "military" finish).

- To teach the finish, the dog is lured with food or the ball (in the left hand) from position in front, past the handler's left side, and behind the handler about 2 or 3 feet (the handler normally takes one long step back with his left foot). Then the handler turns the dog in toward himself (the dog turns counter-clockwise), and leads the dog forward into heel position and asks the dog to "Sit." When using food, the handler then allows the dog to eat from the left hand while in heel position. When using the ball, the handler holds the ball in the left hand just in front of the dog's nose, gives the "Yes," and flicks the ball into the dog's mouth. Initially the handler leads the dog through the entire path, covering nearly as much ground as the dog does. With time the handler moves less and less, taking a smaller step backward and shortening the circle he makes with the left hand and the reward. Eventually the handler does not step backward with the left foot, simply commanding the dog to "heel" and making a long gesture with the left hand to send the dog past the left side and behind, and turn the dog and bring it back up into heel position. The amount of room the dog is given to accomplish the movement is gradually decreased, and when the entire teaching sequence is performed well, the dog begins "flipping" or "swinging" to heel position (rather than walking behind the handler, turning around, and walking up into heel position). A wall or barrier is often used initially, to help the dog reach a straight heel position at the conclusion of the movement. If the dog sits crookedly or too far forward, the handler gives the "No" cue and makes the animal repeat the whole exercise correctly before rewarding.
- In the next step the dog's attention is shifted from the left hand to the handler's armpit. When the dog reaches heel position, the handler marks this behavior with "Good," and then carefully moves the left hand and the

ball up above the dog's head to a position just in front of/under the armpit. Then he gives the "Yes" and drops the ball into the dog's mouth.

- After a few repetitions, the handler actually places the ball in his armpit prior to rewarding the dog. The movement begins as before. The dog is led through the finish with the ball in the left hand, concluding with the dog at heel position and the ball held directly in front of the dog's nose or on the left side of the dog's head. The handler then marks the correct completion of the exercise with "Good," and then raises the left hand above the dog's head, transfers the ball to the right hand, and places the ball in the left armpit, clamping it there with the bicep. Then the handler drops both hands to natural positions, with the left as always hanging just outside of the dog's head. The dog should stare straight up toward the ball from heel position. The handler rewards the dog by saying "Yes" (the dog will normally release and rear straight up toward the ball) and delivering the ball directly into the dog's mouth by unclamping the bicep so that the ball drops free.

Reward in position at heel

Eventually, the handler begins leaving the ball in his pocket until the dog correctly finishes. We now expect the dog to complete the finish to obtain the ball reward, but without seeing or following the ball reward. The handler makes the same motion with the left hand, and even cups the hand as though he is holding a ball, but the ball remains in the right pocket (at this point the leash is normally transferred to the left hand for the first time and held in a manner similar to holding the ball). When the dog reaches correct heel position, the handler marks with "Good," and reaches with his right hand into the right pocket, withdraws the ball, reaches across the body and places the ball in the left armpit, and then says "Yes" and drops the ball. This is a type of reward in position technique (see the "Reward in Position" section under "The 'Clear Signals' Part of CST" above).

Another technique for reward in position at heel

In a variation on this procedure, the handler does not place the ball in his armpit after the "Good" cue, but instead places the right hand with the ball directly in front of the dog's head, or to the left of the dog's head (to bend the dog's head away from the handler and straighten the animal's spine, and then gives the "Yes" release and rewards in position. Once the dog can correctly complete this skill, it has learned a specific position and a specific movement that will become the basis of heeling/marching.

"Heel" word alone

Eventually the gesture of the left hand (the prompt) that is used to send the dog to heel is faded out, so that the dog swings into correct heel position and into attention on the word "heel" alone.

Marching at heel position

Now we are ready to teach heeling proper, actual marching with the dog at heel position. Initially the handler keeps the ball in the left armpit to give the dog a focal point. Holding a very, very short leash in the left hand (from 3 to 8 inches, but without any tension on the leash between hand and the dog's collar), with the hand outside of and just behind the dog's head, the handler gives the command "heel" and shuffles very carefully forward a few feet. When the dog moves well, between the handler's left hand and hip while looking straight up at the ball held in the handler's armpit, then the handler marks this behavior with "Good," and comes to a halt carefully so that the dog does not lose position (initial heeling is often performed along a wall or fence). Once the dog is in sit-halt position, then the handler gives the "Yes" release and drops the ball into the dog's mouth. It is important to remember that in the initial stages of marching, the dog is rewarded after only a few steps, and normally always in the sit-halt. That is, when the dog moves well, the handler marks this with "Good" and then comes to a halt and performs reward in position.

Further practice

With further practice the trainer can leave the ball in the right pocket, asking the dog to move at heel while looking up at the handler's face or armpit rather than the ball. Once the dog moves well, then the handler marks this behavior with "Good," comes to a halt, takes the ball out of the pocket with the right hand, and then performs reward in position in one of the two ways described in the "Reward in Position at Heel" and "Another Technique for Reward in Position at Heel" sections above. Place the ball in the left armpit and drop it to the dog after saying "Yes," or transfer it to the left hand and hold it near the dog's head while it is sitting in heel position, and then flicking it into the dog's mouth after saying "Yes." Gradually the dog is taught to heel for longer periods with the ball in sight and without the ball in sight, depending on the circumstances.

Longer heeling pattern

Eventually, the heeling pattern is made longer, with turns and halts and changes of pace, and the dog is made to work for extended periods for the "Good" and the "Yes" and the reward. If, while moving, the dog loses position by running wide or forging forward or swinging out into a crabbing motion, then the handler marks this error with "No," halts, and re-commands the dog to finish to heel. Once the dog is back in correct position, the handler encourages with "Good" and resumes heeling again. If the dog this time maintains correct position, it is given "Good," sit-halt, and reward in position as in "Reward in Position at Heel" and "Another Technique for Reward in Position at Heel" sections above.

- Up until this point, the "Yes" has normally been given only after the "Good," as a way of releasing the dog into the ball. This practice has been advantageous because it served to make sure that the dog always got its reward while holding the desired position. Such rewards in place are optimal for making sure that reward anticipation does not interfere with steadiness in the sit and down and correct position while heeling. However, they do not make full use of the power of the well-conditioned "Yes," to instantaneously identify to the dog, and reward, very specific aspects of performance. But now we are ready to begin using the full potential of the "Yes."

- Once the dog shows that it understands how to finish quickly and efficiently to heel position, maintain focus on the handler by looking up toward the handler's face while maintaining correct heel position, move at heel without losing position, and correct itself back to the proper position when it happens to lose position, then we are ready for the final step, in which the handler begins rewarding directly out of heel with the "Yes" marker. At any moment that the handler judges the dog should be rewarded, either after the finish, while heeling, or after the dog corrects itself back into position in response to the "No," then the handler rewards the dog by saying "Yes." The dog will release from heel and show that it expects reward, and the handler can then withdraw the ball from the pocket and give it to the dog.

- Use of the "Yes" marker in this way enables the handler to bring to the dog's attention and selectively reward very fine-grained aspects of performance, such as small differences in speed of movement, angle, or posture. Competitive trainers find this useful because they are interested in polish and speed and precision, because these are the things that win trophies. However, MWD trainers should also be interested in polish and speed and precision, because these are the hallmarks of a dog that fully understands commands and skills. Only when the dog has full understanding is it fair and effective to apply pressure to the dog to make sure that it always performs correctly, even under real-world conditions where failure to perform is dangerous for dog and handler and those personnel that depend upon the MWD team.

- Note that the "heel" command is given only to finish the dog, and when the handler first goes into motion. The handler may re-command "heel" after a "No" also. In finished form the command for finish and for heeling is the verbal command "heel" only—there is no gesture of hand or body.

Training Heeling

- We begin training the heeling skill when the handler no longer has to hold the ball in the left hand to signal/lure the dog to heel position (see "The Finish" subsection in "Teaching Heeling" above). Now, instead of the ball, the handler holds the leash with no more than 12–18 inches of slack. After saying "heel," the handler delivers a slight popping correction on the leash. As always during the training, the leash input does not force the dog to finish. The dog is already on its way to heel position, because of habit and its desire for the ball. This technique merely serves to "connect" heeling and the leash input, and bring the finish under the control of the leash input. Similarly, slight pops can also be given as the dog arrives at heel position to encourage it to stop in the correct place, as it sits, and while it is sitting to encourage it to keep its eyes focused up at the handler. All of these movements are thus brought under the control of leash inputs, so that subsequently the dog will not be confused if it receives a correction (see "Proofing Heeling" below) for failing to complete any of them quickly enough or correctly.

Proofing Heeling

- Once the finish and heeling, the halt, and attention to the handler have all been trained/"connected" to leash inputs, then we are ready to begin making the "heel" command an obligation for the dog rather than request. This is done by sharpening the leash inputs so that the dog begins to deliberately avoid errors to avoid the corrections. Plenty of ball reward is provided during proofing to keep the dog's drive up and ensure that it still enjoys heeling work. Most corrections are preceded by the marker "No" (given exactly when the dog goes wrong), to make absolutely sure that the dog associates the correction with what it did wrong rather than some other behavior. The handler must realize that if the "No" is used properly to mark undesirable behaviors, the correction need not be delivered immediately. In fact, in many cases the technique works better if the handler is very deliberate and takes his time delivering the correction in a calm and measured manner (as long as the "No" has been delivered at exactly the right moment).

- Corrections for attention—to bring the dog's focus to its handler—are made by lifting the leash-hand quickly directly toward the handler's face. Once the dog comes into proper attention, then the animal receives "Good" and reward in position, or simply the "Yes." Almost all other corrections are made directly toward the dog's tail, jerking backward with the leash held in the left hand, directly behind the dog's head, over its back, with only a few inches of leash held between the hand and the dog's collar. In this way the hand hangs in a natural position just behind and outside of the dog's head, and any input with the leash serves to stop the dog from coming too far forward past the handler's knee. When a dog is taught to heel using the ball (and generating prey/retrieving drive), most of the errors are related to the dog wanting to move too far forward while heeling and come across the handler's body, in anticipation of receiving the ball. The leash corrections just described are ideal for counteracting this tendency and encouraging attention, when combined with proper rewards in position as described above (see "Reward in Position at Heel" section in "Teaching Heel" above).

- Once left turns are begun, the same correction will help to prevent the dog from bumping or riding against the handler. When right turns are begun, if necessary the left hand holding the leash is brought from behind the dog's head across the handler's front and up, to generate a correction that brings the dog's head toward the handler's centerline and up toward his face.

- During the proofing stage, the handler reduces the frequency of ball reward (substituting praise and petting) and asks the dog to heel for longer and longer periods, and with greater and greater distractions. The eventual goal is tactical heeling capability—meaning that the dog finishes to heel from any distance on verbal command, and remains at heel after one command no matter what the handler's pace or direction, off-leash, and under intense distraction (such as gunfire and stimulation from decoys dressed in bite gear cracking whips, etc.).

CONTROLLED AGGRESSION TRAINING WITH CLEAR SIGNALS TRAINING

Definitions

Decoy or agitator
The trainer who plays the role of suspect or aggressor for the dog, and gives the dog bites. The decoy's skill and ability are critical to success in training, accounting for at least 50 percent of the finished product.

Agitation
The art and practice of provoking aggression and biting behavior from working dogs as performed by the decoy or agitator.

Civil agitation
Agitation performed by a decoy or agitator who does not wear any bite equipment; hence "in civil."

Rag
A piece of jute or burlap, often in the form of a feed bag, used to excite and provoke the dog, and allow it to practice biting.

Sleeve
An arm protector worn by the decoy on which the dog bites. In DoD, the sleeve is often referred to as a "wrap," from the days in which bites were given by wrapping the arm with fire hose or similar materials.

Bite-bar sleeve
Hard sleeve made with plastic and/or leather barrel, equipped with upper arm protector, and with a blade-like "bite bar" projecting from the forearm area and meant for the dog to bite.

Soft sleeve
A soft arm protector made of padding and synthetic or jute fabric. Often used to strengthen or "build" the dog's bite for harder sleeves. Also sometimes referred to as a *puppy sleeve*.

Intermediate sleeve
A firm sleeve of padding and synthetic or jute fabric, often patterned after Belgian bite sleeves used for training Ring Sport. In DoD intermediate sleeves have traditionally not been used—instead the bite bar sleeve was emphasized. Clear signals patrol training makes extensive use of intermediate sleeves for several reasons:

- The intermediate sleeve has a better "bite building" effect with many dogs than the bite-bar sleeve.
- The intermediate sleeve is more versatile; appropriate for soft biting dogs through very hard biters; and enabling bites on the upper arm and the insides of the arm.
- The intermediate sleeve is safer for the dog; protecting it against impacts, collisions, and twists that break teeth and injure necks and spines when they occur on hard bite-bar sleeves.

Hidden sleeve
A firm sleeve resembling a very small intermediate sleeve, made so that the hand is exposed, and meant to be worn under the sleeve in a "concealed" fashion. The hidden sleeve is designed to render the biting dog less "equipment dependent."

Bite suit
A heavy, padded suit with a synthetic fabric outer surface on which the dog can bite anywhere on the arms, legs, or body.

Whip

A short-handled whip with a lash that is used by the decoy to create motion and noise (by cracking it) to provoke and excite the dog. Reed sticks and split bamboo batons are used in a similar fashion. All of these instruments may be used to test the dog's nerve and prepare it for combat (use of the whip does not include striking the dog).

Full bite

A manner of biting in which the dog employs its entire mouth, rather than just the front teeth, while biting. In general, a dog that "bites full" is more confident and reliable than a dog that "bites shallow."

Commitment

The habit of biting without hesitation or prudence and with full force. A dog that bites with commitment flings itself into the decoy with impact and shuts its mouth instantly with all of its strength.

Equipment-oriented

The habit or tendency to pull toward, bark at, and try to bite decoys wearing visible bite equipment, or bite equipment itself lying on the ground, rather than the decoy "in civil."

Man-oriented

A dog showing a great deal of man interest.

Man interest

The habit or tendency to pull toward, bark at, and try to bite the (unprotected) decoy "in civil" as opposed to a decoy wearing bite equipment, or bite equipment itself lying on the ground. Can also be called "civil aggression," but man interest is a better general term for a dog that tries to close with unprotected agitators, whether it does so because it is hunting them (prey drive) or because it likes to fight (active aggression or dominance drive) or because it has the habit of offensively defending itself when provoked (defense drive).

Transfer

A skill in which a dog voluntarily releases a piece of bite equipment that the decoy drops, and redirects its attention to the decoy. The dog may transfer because it is a very civil aggressive animal (for which the transfer comes very easily, because the dog's man interest is strong) or because it has been carefully taught to do so. Transfer is used to denote the shift of attention only, not an ensuing bite. Thus, when we say "transfer the dog," we mean that the dog is induced to release the bite equipment and shift its focus back to the decoy. A second bite may or may not then be delivered.

Drives

A term used by dog trainers to describe the intensity and the quality of a dog's goal-motivated behavior.

- *Prey drive*. The motivation said to cause the dog to search for, chase, and bite objects (including people) that "remind it" of prey animals like rabbits. In prey drive, the dog is relatively unstressed. It seems to enjoy itself and does not growl or snarl or show its teeth—it merely chases or approaches and bites. Prey drive is associated with full-mouth biting. The very prey-oriented dog tends to be very equipment-oriented and does not transfer easily from equipment.

- *Defense drive*. The motivation said to cause the dog to defend itself aggressively from other animals (including people) that threaten or frighten it. When behaving defensively, the dog is stressed, it does not appear to enjoy itself and it growls, snarls, and displays its teeth prominently. A very defensive dog exhibits pronounced signs of fear and stress. The defensive dog is normally very man-oriented and transfers easily from equipment.

- *Fighting drive*. The motivation said to cause the dog to perform work (like search for prolonged periods) to close with, and fight, a person. The behavioral signals said to indicate fighting drive are somewhat indistinct and poorly defined. The term is used mainly to denote a dog that combines characteristics of prey and defense. The animal works with the intensity and man interest typical of self-defense, and appears to be very man-oriented, but does not exhibit the stress and fear typical of a very defensive dog. Also referred to as active aggression, and sometimes as dominance drive.

Nerves

A term used to describe the degree of emotional stability or calmness the dog appears to show while engaged in bite work. A very nervous dog appears anxious and stressed while working, is prone to snarl and growl, and tends to bite with a small or shifting mouth. A very steady or "clear-headed" dog appears un-stressed while working (although perhaps very excited), is not prone to snarl and growl, and tends to bite with a full mouth without shifting. Dog trainers express these ideas with remarks like "the dog is 'nervy,'" for undesirable behavior, or "the dog has good nerves," or "is clear-headed," for desirable behavior. For dogs that are extremely stressed while biting, normally as a result of excessive or poorly applied compulsion, the term (from German) "hectic" is frequently used.

Flat collar
A flat, buckled collar made of nylon webbing or leather and meant for the dog to pull against comfortably. The flat collar should fit loosely and sit low on the dog's neck to provide for comfortable pulling without choking.

Correction collar
A choke collar made of light chain or of nylon cord, or a pinch collar. The correction collar should fit snugly and ride high on the dog's neck above the flat collar.

Back-tie
A technique, and the line used for it, in which the dog is anchored by means of a line or rope attached to a secure point (fence or post, or bolt anchored in a wall) and clipped to the dog's flat collar. Depending on the need and the situation, the back-tie is often equipped with an elastic section made of bungee cord or bicycle inner tube that allows the back-tie to stretch and give a few inches, encouraging the dog to pull, and protecting its spine against shocks.

Basic Bite Work
MWDs are selected for DoD purchase by means of a consignment test. The patrol portion of this test emphasizes the dog's willingness and ability to defend itself (i.e., defense drive) rather than its raw desire to engage in bite work (prey drive or active aggression). Training for patrol certification and also most types of patrol MWD utilization/deployment (e.g.,

scouting, building search, and pursuits) require that the dog enjoy bite work (either through hunting/prey behavior, or because the dog likes to fight—active aggression) rather than just be self-defensive. Accordingly, the first order of business with a "green" MWD is basic bite work, in which the dog's desire to bite, physical condition and power, and bite-targeting skills are developed. In addition, either in the course of basic training or in the course of more advanced training in the field, the dog must learn to bite any area of the decoy's body through training on the bite suit, and also learn to bite concealed/hidden sleeves. Exceptionally strong dogs also benefit from attack work on "civil" decoys in the agitation muzzle. The objectives for basic bite work are:

1. The dog should confront, with barking and lunging and attempts to bite, an agitator "in civil" that approaches and threatens the animal.
2. The dog should bite with commitment, power, and as full a mouth as possible on intermediate sleeves, and if possible on hard sleeves.
3. The dog should transfer (release bite equipment when the decoy drops it), and attempt to approach and bite the decoy in preference to the equipment.
4. The dog should continue to bite, without excessive growling or shifting of the bite, when threatened and struck with a whip/stick.
5. The dog should pursue a decoy at full speed over a distance of at least 50 yards, bite with commitment and power, and continue to bite without disturbance as the handler approaches and takes the leash and praises it.
6. The dog should perform all of the above skills indoors on slick surfaces as well as outdoors.

Basic Bite Work Session
All of the above objectives are achieved through training sessions resembling the following:

- The dog wears two collars—A flat leather nylon collar and a correction collar, and is attached to a short back-tie anchored well above its back and 5 to 8 feet long. The handler stands near the dog holding the leash (attached to the correction collar).The decoy stands out of sight behind some obstacle, "in civil" with or without whip/stick.

- The session begins with the "Watch 'Em" command from the handler, then the decoy steps into view and begins to work, provoking and exciting the dog with threats and aggressive postures and movements. Initially, the dog may have very little reaction when hearing the "Watch 'Em" cue, but after a few sessions, the dog will learn the association between "Watch 'Em" and the appearance of the decoy (classical or Pavlovian conditioning) and it will become excited and aggressive on command. This is called "alerting" the dog.

- In real-world law enforcement and military patrol dog applications, the alert is critically important, because the dog's aggression and biting must be under the control of the handler's command rather than under the control of the decoy's appearance or behavior. A high percentage of patrol dog deployments in which the dog is called upon to engage/bite personnel involve passive subjects, or subjects that may not be moving as vigorously or shouting as loudly as non-target personnel in the area. Accordingly, the handler must have the capability of cuing the dog's drive, so that it wants to bite, and then telling it whom to bite.

- The decoy works by alternately threatening the dog (by moving directly at the animal, staring into its eyes, pretending to hit or strike at it, and vocalizing angrily), and by yielding to the dog (turning away from the dog, taking a step back, running away, pretending to be afraid). The object of the exercise is to make the dog more resistant and confident in the face of threats, and therefore the decoy must yield when the dog reacts powerfully to threats (i.e., counter-threatens) by lunging forward, barking, or attempting to bite. The handler's role is to encourage the dog (not too loudly) and praise it when the decoy runs away, but the handler should not intrude too much into the situation. The primary trainer in this situation is the decoy and we must let the dog concentrate on him.

- The decoy has another means of relieving stress, which is to "channel" the dog's energy and emotion from aggression (defensive or active) into prey. The decoy accomplishes this channeling by reacting to a counter-threat with rapid, exciting lateral movement. This "rabbit-like" motion stimulates prey impulses, brings the dog forward offensively, and "unloads" stress.

- After a brief passage of civil agitation, the decoy retreats to the hiding place and puts on a pair of intermediate sleeves, one on each arm (see "Definitions" note above about intermediate sleeves versus bite-bar sleeves). The handler again alerts the dog with "Watch 'Em," and the decoy appears and re-agitates the dog. This time, when the dog counter-threatens by lunging and/or barking, the decoy delivers a bite on one of the intermediate sleeves. If the dog is very powerful and appears confident, then the decoy can deliver the bite by moving directly into the dog. If the dog appears less powerful and confident, then the decoy moves laterally, coming close enough for a bite by zig-zagging right and left, approaching diagonally.

- Note that the quality of the dog's behavior is likely to change noticeably from aggression (defensive or active) to predatory when the decoy wears sleeves, and the result may be that the dog becomes less sensitive and reactive in response to threats from the decoy, instead merely lunging toward the decoy in efforts to engage the equipment.

- The bite is delivered by holding the arm high and across the chest or upper abdomen, and encouraging the dog to jump up and strike the sleeve, rather than by swinging the arm into the dog's mouth. Once the dog bites, the agitator struggles with the dog and yells and vocalizes, always being careful not to overwhelm or frighten the dog. If the dog takes a shallow bite or shifts its bite nervously, then the decoy puts tension on the back-tie by leaning backward, threatening the dog with having the sleeve pulled from its mouth. This should cause the dog to bite harder and, if the agitator suddenly reduces tension on the back-tie and pauses for a moment, it may cause the dog to bite in and take a fuller bite. This act of "biting-in" is often rewarded by resuming movement, or "yielding" to the dog by pretending to stagger or fall back. Sometimes the "bite-in" is rewarded by letting the dog take the sleeve off of the decoy's arm.

- When the agitator allows the dog to pull the sleeve off of the arm, he steps back out of range of the dog's bite, and immediately begins to agitate the dog. The dog should lose interest in the sleeve and transfer, dropping the loose sleeve and lunging to bite the decoy again. Once the dog transfers, then the decoy delivers a bite on the other intermediate sleeve.

- If the dog does not transfer voluntarily, then the handler makes the dog release the sleeve by lifting up on the collar(s). During this process, the decoy stands passive. Once the dog has released the sleeve, then the decoy agitates and delivers the next bite.

- While struggling with the dog on the second bite, the agitator uses the free hand to pick up the first sleeve and put it back on. At a moment when the dog is biting well he releases the second sleeve, transfers the dog again, and gives another bite on the first sleeve. The session proceeds like this for three to six bites. On the last transfer, the decoy does not give a bite, instead he attracts the dog to the side, away from the grounded sleeve, the handler gets control of the dog, and the decoy runs away with the sleeve. The dog is left victorious, having bitten several times, transferring each time, and finally chasing the decoy away.

- This basic session is designed to: a) teach the dog to alert powerfully on command (by lunging and barking when given the "Watch 'Em" cue, even though it cannot yet see the agitator), b) strengthen the dog's man interest, c) build the bite, and d) teach the dog to drop "dead" training equipment and shift its attention back to the decoy (transfer).

Intermediate Bite Work Session

The intermediate bite work session progresses much like the basic session, except in three respects: a) The decoy conditions the dog to withstand and fight back against stick-threats; b) The dog is expected to transfer to the decoy in civil; and c) The session end with off-leash pursuit bites.

- Counters to stick-threats. During the bites, the decoy begins to threaten the dog more vigorously. The decoy does this by looking strongly into the dog's eyes and moving the hands and the stick/whip sharply at the dog's face and body without striking the dog. Initially these threats are relatively weak, but as training proceeds they become more violent and longer in duration, finally concluding in a simulated fight between dog and decoy. If these threats are performed correctly, the dog does not avoid, or "back off" of its bite, instead it "counters" powerfully by "biting in" and pulling and head-shaking. The decoy rewards these "counters" by yielding, and sometimes letting the dog take the sleeve. The result is a dog that knows how to fight a person, and feels confident that it can win the fight, even when the decoy exerts considerable psychological and physical pressure.

- The handler's role during this process is to encourage the dog, but not so loudly that he distracts or disturbs the animal. At least 90 percent of the responsibility for training at this point is the decoy's.

- Because serious fighting, even victorious fighting, causes accumulated stress, the trainers must not pressure the dog during every bite work session. Some sessions are easy, even fun; some sessions are slightly more serious; and very rarely the trainers design a session to test and strengthen the dog's nerve.

- Transfer to the decoy in civil. When the decoy drops the first sleeve and transfers the dog, and then allows the animal to bite the second sleeve, he does not put on the first sleeve again. Instead the agitator leaves the first sleeve on the ground, and then releases the second sleeve to the dog and steps back out of range (i.e., the decoy is now "in civil"). By now the dog should have learned a very automatic transfer. The moment the dog recognizes that the decoy has dropped the sleeve, it should release the sleeve and redirect its focus to the decoy. In this case, the dog is transferring to the "man" rather than to the equipment. The moment the dog transfers, the decoy rewards the dog for this behavior. How the decoy accomplishes this reward depends upon the type of dog.

- If the dog has very high man interest (presumably because it is very high in defense drive or fighting drive), then all the decoy needs to do is to react to the dog by vocalizing and moving vigorously, and then running away. A defensive dog will be gratified because it has chased away the enemy that is causing it stress. An actively aggressive dog (with abundant fighting drive) will be reinforced because it has won possession of the battleground and increased its sense of dominance. Of course, if either of these types of dog have ample prey drive as well, when the decoy runs away this rapid movement will also stimulate hunting behavior, which is reinforcing for the dog.

- If the dog has a higher degree of equipment orientation, so that it tends to be reluctant to transfer from the dead sleeve, or it tends to return to the sleeve after releasing it (presumably because it is a dog in which prey drive predominates), then the decoy must provide a bite reward. If this type of dog does not receive a bite reward of some sort when it transfers to the civil agitator, then it will be "disappointed," or punished (by omission of the bite) for the transfer, and soon it will stop transferring to the civil decoy and continue biting the sleeve when the decoy drops it.

- In the case of a very strong-biting dog of this type, the decoy can provide this bite on an arm- or leg-sleeve hidden under the outer clothes. The opportunity to transfer to a decoy that appears to be "in civil," and then bite that decoy, teaches the dog a very important lesson—the man *is* prey. At the conclusion of the bite the decoy drops prone, and the handler removes the dog from the bite physically by lifting upward on the collar(s), and then the decoy runs away.

- In the case of a less powerful dog that may not have the necessary drive and confidence to bite the hidden sleeve, the decoy can pull a jute rag from its hiding place in the waistband at the small of the back and let the dog suddenly bite it. This is not as powerful a technique for producing man interest in a prey/equipment-oriented dog, but it does reward the dog for directing its energy and attention at a civil decoy. After a few seconds of biting and a vigorous fight with the rag, the decoy releases the rag to the dog, transfers the dog from this rag (usually by picking up one of the sleeves), and runs away.

- Pursuit bites. Following a session of bites and transfers to the decoy as above, the decoy runs away "in civil" and picks up two more intermediate sleeves lying on the ground at the desired distance for the pursuit bite. The handler unhooks the dog from the back-tie and releases the dog to pursue and bite. The decoy takes this bite and keeps the dog occupied with fighting while the handler runs up, being careful not to disturb or frighten the dog, and praises and pets it enthusiastically while it bites. Once the handler has the leash, then the decoy drops the first sleeve and steps back out of range. The dog will transfer, and then the decoy runs back toward the original back-tie location. Again, the handler releases the dog for a pursuit bite, follows the dog up and praises it and regains the leash. The session can end in one of two ways:

  1. If the dog has very high man interest and transfers easily to a civil decoy, ignoring the sleeve after the transfer, then the decoy drops the sleeve, the dog transfers, and the decoy runs away with the dog, restrained by the leash, in hot pursuit. The handler gradually brings the dog to a stop, and the decoy escapes.

  2. If the dog has lesser man interest, so that it is not clear that the dog prefers the decoy to equipment, then during the bite the decoy picks up and puts on one of the original sleeves. Then he drops the sleeve the dog is biting, transfers the dog by letting it see the sleeve he is wearing, and then the decoy runs away with the dog, restrained by the leash, in hot pursuit. The handler gradually brings the dog to a stop, and the decoy escapes.

Controlled Aggression

There are two crucial points in the career of any practical patrol K9 or MWD. One of these, of course, is when the dog is first called upon to engage a subject. The other is when the handler for the very first time begins to exert control over the dog during bite work. When we say "control" we mean verbal commands enforced with physical corrections. Depending upon how this stage of the dog's training is performed, it is more or less stressful for the dog, and it is the first "acid-test" of the dog's quality. Many dogs cannot withstand the stress and do not satisfactorily evolve into well-controlled but hard-biting patrol dogs. Therefore, as in any phase of CST, the trainer's main concern is minimizing the degree of stress the dog suffers by ensuring that the animal understands the skills it is taught, and as much as possible avoiding the use of any substantial corrections until it is clear that the dog understands the target skill and is fully capable of executing the skill. However, in controlled aggression training (as opposed to obedience), because it is often difficult to with-hold the reward from the dog to punish it (i.e., prevent it from biting), we are necessarily more dependent upon physical correction and compulsive training. Therefore the emphasis in controlled aggression training is on the *training phase* (in which the dog learns what corrections "mean" and how to handle them) and the *proofing phase* (in which the dog learns that it must obey certain commands or it will receive meaningful correction).

- Out and guard versus recall. The modern DoD patrol dog is trained to meet an "out and guard" standard. This means that the dog is trained to release the bite on command (or "out") and then remain near the decoy, guarding intently (silently or while barking), as opposed to releasing the bite and returning to heel position as was taught during the era of "six phases of controlled aggression." Similarly, when the dog is called out prior to the bite, while it is running downfield after the decoy (i.e., the "stand-off"), it is supposed to halt and remain guarding the decoy. In contrast, the "six phases" dog was taught to return to heel position after being called out during the standoff.

- The out and guard approach is technically more sound from a training standpoint than the out and recall approach. It is a better platform for teaching the dog to understand the out, and adapt to being controlled by the handler during bite work. However, from a liability point of view, it is inadvisable to have a law enforcement K9, or an MWD used in a law enforcement role, guard a suspect/subject closely, because of the risk and likelihood of unwarranted bites. Similarly, from a tactical point of view, it is extremely dangerous for the handler to have the dog guard the subject unless the handler has the ability to recall the dog to heel from the guard position, because to recover the dog, the handler must approach the subject while the dog guards, and potentially leave a position of cover for an exposed position to do so.

- Therefore, at some point in its career and training, the out and guard patrol MWD should be taught to recall to the handler from the guard, over any distance at which the dog might be sent to engage a suspect.

- Out and guard standard. To be certified as an out and guard patrol MWD, the dog must perform the following exercises:

    1. *Field interview.* In which the dog remains at heel and under control while the decoy approaches the handler, converses with him, and then departs at a walk.

    2. *Attack.* In which the dog pursues the decoy on command, bites and holds, outs on command, and then guards while the handler approaches and places himself at heel position.

    3. *Search and escort.* In which the dog guards the decoy while the handler searches him, and then walks under control with the handler while escorting the decoy back to the starting point.

    4. *Search and re-attack.* In which the dog bites the decoy without being commanded when the decoy attacks the handler during the search.

    5. *Standoff.* In which the dog is called out in mid-pursuit and stops in a standing, sitting, or lying position and guards the decoy while the handler approaches and places himself at heel position.

    6. The essential aspects of all these skills are taught to the dog in the course of sessions in which the dog is back-tied, with the back-tie attached to the flat collar and the handler holding the correction collar (see "Definitions" in "Controlled Aggression Training with Clear Signals Training" above).

- Communication during clear signals controlled aggression training. If anything, use of clear communication cues is even more important in controlled aggression than it is in obedience, because the dog is intensely excited and therefore easily confused, and because controlled aggression can be extremely stressful for the dog if it becomes confused about what the trainers want and how to obtain reward and avoid punishment. Fundamentally, the same cues/response markers are used in controlled aggression training as in obedience:

1. "Yes" signals to the dog that it may release from control and bite the decoy, or try to bite the decoy by pulling against the leash. Often in bite work a trainer will use a phrase like "Get 'Em" instead of "Yes," but the phrase is conditioned and used in controlled aggression exactly as the "Yes" is used in obedience, as a terminal bridge (in addition to an alert cue, see "Basic Bite Work Session" above).

2. In controlled aggression, the bite can be signaled by the handler with "Yes"/"Get 'Em," or by having the decoy move so as to deliver a bite (i.e., simulating either an attack on the dog, or an attempt to escape). Which approach is used depends on the dog and the specific situation, but as a rule very powerful dogs that are difficult to control receive their bites after a "Yes" from the handler (so that they think about their handlers, and their responsibilities to their handlers, all the time during controlled aggression), whereas "softer" dogs, that do not guard as powerfully and worry continually about where their handlers are, receive their bites after a movement by the decoy (so that they think about the decoy all the time and as a result guard more intently).

3. "Good"—Intermediate marker that signals to the dog that it is performing correctly and it will be rewarded with a bite or a chase, but it must continue to perform until released.

4. "No" signals to the dog that it has made a mistake and it will be punished. The punishment can take two forms: negative punishment (or omission) in which we withhold reward (the bite) from the dog, and positive punishment, in which we apply some more or less uncomfortable input to the dog (usually a collar correction). The "No" also tends to be a terminal marker in the sense that any time the dog hears a "No," it means that the animal has made a mistake and the last exercise will have to be repeated.

5. The clear signals "No" versus the traditional DoD "No." This manner of employing the "No" is utterly different than the traditional DoD method of using "No." Typically in DoD the handler was taught

to say "No" (perhaps many times while giving a continuous correction) and correct the dog simultaneously. This makes the "No" nothing more than part of the punishment, and it gives the dog very little information. As it is used in CST the "No" is a much more powerful tool that enables the handler to identify for the dog exactly what it did to "get in trouble" and, if the handler allows, self-correct to "get out of trouble."

6. Out (release bite on command). The out is seemingly the simplest of skills for the dog to execute—simply opening the mouth. But, in releasing the decoy, the dog is relinquishing its grip on the single most motivating object it knows. This means that tremendous psychological currents and emotional turbulence can be caused by the out, especially when strong physical force is used on the dog to accomplish the release. In short, although the out is simple, it is far from easy for the dog, and the better and more powerful the dog, the more difficult it can be for the animal to learn to control itself sufficiently to obey commands. This means that the trainer is under a tremendous obligation to be patient, skilled, fair, and humane with the dog in this phase of training. CST methods assist us to meet this obligation by giving us powerful tools for showing the dog what we want it to do, and also by giving us a hierarchical method of adjusting the dog's level of motivation, and hence the difficulty of the exercise for the animal.

• Hierarchy of bite objects and motivation. To adjust the dog's level of motivation, the trainer chooses from an assortment of possible bite objects, ranging from very "unappealing" objects like hard wood and plastic, to more motivating objects like rubber hoses and tug toys, to the most motivating bite objects such as decoys wearing sleeves and suits. This range of objects, and the range of motivation they produce—from low to high—is called a "hierarchy." The basic approach to instructing the out in CST is to manipulate this hierarchy so that the dog learns the skill in a situation of the lowest possible stress (using a bite object that the dog "likes" enough to bite but not enough to be stubborn about keeping a hold on it), and then gradually moving up the hierarchy to the target situation in which the dog is fighting and biting a decoy.

- Teaching/Training the out. Because many "green" dogs procured by DoD already have extensive experience at the time of procurement in being choked off of bites with choke collars, and struggling against handlers who are trying to physically control them, often it is not possible to persuade these dogs to perform an entirely inductive, or voluntary, out. Such dogs have already been taught to be resistant and anxious while biting. Therefore, in many cases some sort of physical input or correction is needed to cause the animal to release on command, so that it can be rewarded with another bite. For this reason, because corrections may be involved in even the earliest stages of instructing the dog to out, teaching and training stages are often not distinct, and they are here discussed as one.

1. The out is best taught during play with the handler (low on the hierarchy of motivation), rather than during bite work on a decoy (extremely high on the hierarchy of motivation), because during bite work the dog becomes extremely excited and therefore may require heavy (and stressful, even potentially injurious) corrections, *unless it has already been taught at a lower level of the hierarchy how to release the bite on command for a reward.*

2. An object is chosen that generates low to moderate motivation. For a somewhat "low-drive" dog, this might be a tug toy; while for a high-drive or "problem" dog, it might be an object like a piece of hard wood or plastic. In the case of the latter type of dog, the goal is to find an object the dog "likes" enough to bite, but not enough to cause it to stubbornly fight the handler to retain the object.

3. The handler entices the dog with the bite object, holding it in both hands by the ends, allows the dog to bite it, and then provides a brief "fight" with praise. Then the handler freezes, holding the bite object still, with the leash (with some slack) held in the left hand along with the end of the object.

4. The handler gives the "Out" command, and then causes the dog to release. Ideally, this is done very gently, either by waiting for the dog to become frustrated by the handler's refusal to play tug of war, or by crowding the dog's mouth off of the bite object with hands and fingers. When neither of these options is practical, a correction is applied, as lightly as possible, by leaning backward

against the back-tie and exerting tension on the correction collar with the leash held in the left hand. Once the dog releases, it is told "Good," and a moment later "Yes"/"Get 'Em" and enticed to bite the object again.

5. The exercise is practiced several times during a session. On the last out of the session, the dog is rewarded by being allowed to take the bite object from the handler (equivalent to the decoy dropping a bite sleeve). The handler then entices the dog with a second bite object, induces a transfer and a bite, gives the dog that bite object, picks up the first object, and transfers the dog back to the first object again. After a few rewarding transfers, the handler ends the session by using the object in his hand to entice the dog to the side, away from the object lying on the ground (that the dog has just transferred from), throws the bite object behind him to keep the dog's attention away from the other object lying on the ground, and then steps in and regains control of the dog with the leash. In this way the session ends "with drive" and with pursuit activity, rather than ending with an out and guard followed by no reward bite.

6. The emphasis here is not on trying to cause the dog to out cleanly on command—it is on teaching the dog to release calmly and willingly to escape a collar correction, then to guard (rather than back away or otherwise avoid) and confidently re-bite when rewarded with the "Yes"/"Get 'Em" cue. At this point we are not especially concerned if the dog ignores the "Out" command and waits for the correction to be applied before it releases. However, typically this stage of training, in which the dog is allowed to practice escaping corrections rather than avoiding them is relatively brief—one or two sessions for a powerful dog that has unshakeable persistence under correction, and several sessions for a softer dog that is more easily upset and put into avoidance behavior.

- Proofing the out. To proof the out, we must drive the dog from escape responding, in which it tends to wait until it feels the correction before it releases the bite object, to avoidance responding, in which it releases on command to avoid the correction entirely. This is done by: a) moving up one or two levels on the hierarchy of motivation so that the dog begins to disobey the out command reliably, and b) applying a sharper correction.

1. Once the dog has become skilled and comfortable in releasing the bite to escape/avoid correction, at a low level on the hierarchy of motivation, we increase the motivation by choosing another object for the dog to bite a little higher on the hierarchy. For a very compliant dog, we may move directly to the decoy and bite sleeve. For a more powerful and resistant dog, we may continue to practice the out only during play with the handler, and merely move one small step up the hierarchy of objects, for instance from a PVC pipe to a firm rubber hose.

2. In either case, if the dog fails to release on the "Out" command, the handler applies a sharp correction with the leash and collar, pulling toward him so that the dog's mouth is pulled "into" the bite object. When the dog releases under this correction, the handler gives the "Good" cue (which now takes on the meaning of a "safety signal" telling the dog that discomfort is over and will not reoccur so long as it continues to obey). After a moment of stable guarding, the handler (or decoy, if we are training on the decoy) delivers another bite and immediately freezes, and the "Out" command is given again. If the dog releases cleanly on command, the handler says "Good" (the safety signal reassures the dog, telling it that it has done the right thing and the danger of a correction is past), and then "Yes"/"Get 'Em," and gives the dog a rewarding bite.

3. In early training, the handler does not say "No" before correcting. The dog simply receives a rapid correction after the "Out" if it does not release. In later training, especially when the handler is at some distance and is unable to deliver a correction immediately (so there will be some delay between disobedience of the command and the correction), then the handler marks disobedience of the "Out" command (continuing to bite) with the "No." Then the handler approaches the dog and calmly and methodically administers the correction.

4. The "Good" cue as a safety signal. When we introduce corrections to obedience and patrol training, the "Good" takes on two meanings: It now tells the dog that no correction is coming, that by

choosing the right behavior it avoided correction; and it also predicts for the dog that, if it continues to perform well, reward in the form of "Yes"/"Get 'Em" is on the way. The purpose of this safety signal is to reduce the dog's stress and anxiety—instead of waiting to see if it will be corrected, it learns immediately that it has found safety from discomfort, and thereby it also learns *exactly which behavior* earned that safety.

5. The dog is taught not to re-bite until the handler signals reward with the "Yes"/"Get 'Em" release. If the dog attempts to re-bite without permission, the handler gives the "No." After the "No," in most cases the handler proceeds to calmly and methodically deliver the correction. However, at some points in training it is advisable to omit the correction if the dog exhibits a very respectful response to the "No" by recoiling from the bite. If the dog self-corrects in this way and the handler decides not to correct, he then gives the "Good" cue (signaling safety for the dog), pauses a few moments, and then the dog is rewarded.

- Guarding the decoy. Once the dog has learned to reliably out from the decoy on command, then we begin the process of developing the dog's skill and persistence in guarding the decoy. This guarding will be the foundation for nearly all the exercises of controlled aggression such as field interview, search, standoff, and so forth.

Teaching/training the dog to guard
The dog is still trained on the back-tie, and it learns the first stages of these exercises while working on the back-tie.

1. In DoD the most frequent cause of severe training problems in patrol, and eliminations from patrol training, is failure to guard—the dog leaves the decoy when it should be guarding, waiting for its next opportunity to bite. This problem is most often caused by faulty training, by handlers pressuring and correcting dogs while guarding until the animal's drive is overcome by its anxiety, and it avoids the situation. This behavior is not disobedience, it is avoidance behavior fueled by anxiety and confusion. To guard well, the dog must be confident that it knows

how to guard without biting, confident that it knows when a re-bite will be "safe" because it is authorized, and it must trust its handler. Accordingly, CST takes a great deal of trouble teaching the dog to guard while still on the back-tie, by breaking the guarding into steps.

2. At all times, it is the decoy who causes the dog to guard by enticing the dog subtly and delivering bites at the right times. The handler does not make the dog guard by pressuring it to "Stay!" However, if while guarding, the dog re-bites without permission, then the handler may deliver the "No" and correction.

3. Guard during handler movement. The first step is to teach the dog to continue guarding while the handler moves. While standing near the dog, holding the leash, the handler takes one step away from the animal. The dog should continue to guard, and then the decoy rewards this behavior by delivering a bite. After the next out the handler takes two steps, and the dog receives a bite, and so on. Eventually we condition the dog to guard intently while the handler walks and even runs about—around the decoy, to the dog and away, behind the dog, etc. If the dog becomes distracted for a moment by the handler and looks away from the decoy, or tries to leave the decoy, then the decoy punishes the animal by jumping away out of range and agitating the dog to produce frustration. If the dog becomes unsettled by handler movement and bites the decoy prematurely, then it receives a "No" and perhaps a correction. If it recoils obediently from the bite on hearing the "No," then the handler may elect to reassure it with "Good" and continue the exercise.

4. Guard during decoy movement. In the second phase of teaching guarding, the handler steps to heel position on the guarding dog, tells the animal calmly to "Stay" in a reassuring but forceful voice, and uses his right arm to wave or push the decoy back one small step. This must be done carefully to avoid triggering a premature bite. When the dog allows this step away and continues to quietly guard, the handler rewards with the "Good" marker, and then waves the decoy back close to the dog. Once the decoy has stepped up close, with the dog still guarding, then the animal is rewarded with a bite (either after a "Yes"/"Get 'Em" from the

handler or after a sharp movement from the decoy, depending on whether we want to emphasize the dog's attention to the decoy or to the handler).

5. If at any point the dog attempts to re-bite prematurely or move away from the decoy, these errors are handled as above—re-bite is met with "No" from the handler, while avoidance is met by an escape from the decoy.

6. With additional training the dog is taught to remain in a stable guard position on the back-tie while the decoy steps backward, turns, and walks up to 100 feet away; or steps backward and turns and runs in place; or steps backward and then turns and runs rapidly away; or steps backward and stands while the handler searches him. (Note that the only situation in which the dog is allowed to bite without command is when the decoy attacks the dog or the handler. If the dog merely sees a person running, this is not authorization to bite.)

7. Reinforcing guarding with reward in position. When the dog obeys the handler by guarding intently while the decoy makes the above-described movements, the handler marks this compliance with "Good" and then signals the decoy to approach the dog very closely and stand, and then give the reward bite. The reward bite may be delivered after a "Yes"/"Get 'Em" from the handler, or in response to the decoy attacking (moving suddenly at) the dog or the handler. In the early stages of teaching the dog to guard, the exercise does not conclude with the dog running downfield and biting; it normally ends with dog being rewarded in the original place on the back-tie where the animal was given the "Out" command, released the bite, and began to (and continues) to guard. The alert reader will recognize that this is a "reward in position" procedure, identical to the procedure used during obedience to help the dog understand how to remain in place without anticipating or breaking position.

8. Reinforcing the dog for vigilant guarding with a reward out of (i.e., breaking from) position. If the trainer judges it advisable to bring up the dog's intensity or attention to the decoy, the dog can also be given a bite occasionally at any point in any of the

exercises by having the decoy suddenly attack or threaten the dog or handler. In response the dog will release from guarding and lunge to the limit of the back-tie (which has a rubber inner tube or bungee cord insert to absorb shock). The decoy can move quickly to the dog and deliver a bite. This is how the dog learns to bite without command when the decoy attacks the handler during a search.

9. Punishing the dog for inattention or looking/moving away from the decoy. If the dog becomes anxious or distracted during any of the guarding exercises, the decoy can punish the dog (through negative punishment, or omission of reward) by suddenly becoming aggressive (but without letting the dog bite), running away to a hiding place, and then pausing for 30 seconds or a minute before coming back and resuming the exercise. After a few such escapes followed by a frustrating "time out" period, the dog will guard more intently.

10. From this basic out and guard exercise performed on the back-tie, with rewards delivered in guard position, we develop: a) the field interview, b) the "false run" (in which the dog remains under control at the handler's side while the decoy walks or runs away), c) the search of the decoy by the handler while the dog guards, and d) the re-attack during the search. Most important, we teach the dog to guard confidently and calmly while the handler approaches and places himself at heel. To increase the dog's confidence and steadiness during handler approach, the handler often gives the "Yes" after arriving at heel, or the decoy delivers a bite without "Yes" by attacking dog or handler.

11. At the end of the session, the dog is normally unhooked from the back-tie and allowed to perform one or two pursuit bites concluding with transfers. However, these bites are not practiced from the guard. The animal is not released (given the "Yes"/"Get 'Em" command) from the guard to run downfield and bite, because these "thrown attacks" develop tremendous anticipation for the run and make the dog difficult to control in the guard position. Before the handler begins sending the dog downfield from guard position, we must first do extensive work to make the dog steady at heel, and teach it to guard quietly, without barking, whining, bouncing, or

lunging, even when the decoy moves off to a distance of 25 or 50 yards. Therefore, all bites from the guard are practiced by having the decoy return to the dog (reward in position). If the dog is turned loose for a pursuit, this is not done from the guard. Instead the handler holds the dog by the flat collar (the dog is encouraged and allowed to lunge and bark so that it knows that it is not "under control"), the decoy runs away, and the dog is released.

Proofing the guarding exercises

1. All the exercises involving guarding, whether the decoy is nearby or at a distance, are proofed after extensive and patient teaching/training so that the dog becomes very confident and stable in guard position and very accustomed to the "No" and appropriate collar corrections. This way the dog will not be tempted to avoid the decoy when corrected during a guarding exercise, but instead will resume guarding confidently after it makes a mistake.
2. Proofing is accomplished by increasing the intensity of correction so that the dog begins to avoid mistakes, so as to avoid corrections. If the dog makes a mistake (i.e., breaks guard position or attempts to re-bite without permission) when the handler is at a distance, then the handler marks the misbehavior with a "No" and approaches the dog calmly to correct. If the dog has learned what "No" means, and has been given plenty of practice to understand how to correctly guard, then this procedure will make it afraid of making a mistake again, not make it afraid of its handler approaching.

Teaching/training the dog to perform the standoff

When the standoff is correctly taught and performed, the dog drops into a down in mid-flight on the command "Out, Down" and guards from this remote position. The dog should not follow up after the out command—that is, it should not run all the way to the decoy and then guard him from nearby.

A good standoff depends on recruiting the dog's cooperation with the "Yes" command, so that the dog is eager to lie down to get its bite. If we begin the standoff exercise by forcing the dog to lie down, it will be much more difficult to teach.

1. The first step is to teach the dog to earn the bite by lying down. In essence we are teaching the dog to change positions on command while guarding, to be "paid." The exercise begins on the back-tie as always, with the dog guarding in sitting position and the decoy standing at a distance from 10 to 35 feet. The handler asks the dog to lie down, by giving the "Down" command plus whatever additional gestures (kneeling and tapping the ground in front of the dog, etc.) will help the dog to lie down even though it is very interested in the decoy. Once the dog's elbows touch the ground, the handler marks with "Good." Then the handler signals the decoy to come in and stand very close to the dog, and then the dog receives a bite (reward in position). This exercise is repeated until the dog is trying eagerly to lie down (even before it is commanded) anytime the decoy steps back and stands.

2. In the next phase the exercise is repeated in much the same way, but the dog is not commanded "Down" when in a sitting guard position, but instead when it is on its feet watching the decoy or even pulling somewhat on the back-tie. In this situation the "Out, Down" command begins to interrupt ongoing activity, rather than just moving the dog from sit-guard to down-guard.

3. Then the dog is taken off of the back-tie and held on leash. Ideally, for the first exercises the dog is walking forward toward the decoy while leaning a bite against the leash, or some similar low-intensity activity, rather than pulling against the leash and lunging and barking with all its strength. The handler's commands "Out, Down" which helps the dog to lie down, and then gives the "Good," and brings the decoy in to stand close to the dog and deliver the bite, providing the dog guards calmly from the down. With repeated exercises, the dog is allowed to move faster or become more excited, pulling harder against the leash until given the "Out, Down" command. In this way, the "Out, Down" command begins to take on the power to interrupt the dog as it moves forward vigorously (but not off-leash) toward the decoy. At this point we are ready to begin allowing the dog to run free of restraint toward the decoy.

4. Preventing hesitation. Simultaneously with the work described above (teaching the dog to lie down to earn a bite), we have also

been preparing the dog in another way, to prevent problems with hesitation. The dog is set up at heel position next to its handler, with a decoy about 50 to 60 feet away. The dog is not back-tied. Normally the handler holds a 6-foot leash that he will drop when he gives the "Get 'Em" command. An assistant stands midway between the handler/dog team and the decoy, holding a long line attached to the dog's correction collar. The dog is sent to bite with the command "Get 'Em," but the decoy stands passive (although he will provide slight enticement if necessary to help the dog bite) so that the dog bites on command alone rather than in response to decoy movement. This procedure is repeated until the dog is thoroughly accustomed to the "standoff" setup and shows no hesitation in biting a passive decoy on command.

5. Once the dog downs readily for bite reward even when it is moving vigorously on leash, and when there is no trace of hesitation when the dog is commanded to bite a passive decoy with the assistant standing midway, then we are ready to run the standoff proper. The setup is as in "Preventing Hesitation" above, with handler and decoy 50 to 60 feet apart and an assistant at the midpoint holding a long line attached to the dog's correction collar. The dog is sent with the "Get 'Em" command, but at the halfway point the handler commands "Out, Down." The assistant uses the long line to rather gently check the dog and bring it to a stop. Once the dog is down, the handler gives the "Good" marker, advances to heel position, and then signals the decoy to come forward, stand near the dog, and deliver the bite (either after "Yes" from the handler or by attacking dog or handler, depending on the needs of the dog). Note that again we are using the reward in position command; to steady the dog in the down and teach it that, once it hears the "Out, Down" command, reward will no longer be earned by running forward toward the agitator, but instead through dropping immediately into the down and guarding.

6. In alternating exercises, set up identically, the dog is sent with "Get 'Em," but "Out, Down" is not given. The dog is allowed to continue and bite the passive decoy. The "mixture," or "balance,"

of bite trials versus standoff trials is arranged to keep the dog from hesitating (i.e., anticipating the "Out, Down" command and running out slowly when told to "Get 'Em"). The decoy never varies his behavior, always standing passive until the dog either bites or is called "Out, Down."

7. Soon the dog should be lying down quickly and eagerly once checked, although it may be necessary for the assistant to use the line to check/stop the dog's forward movement.

Proofing the standoff

1. Proofing is performed by making the checking correction sharper and more uncomfortable, so that the dog begins to check on its own when commanded "Out, Down" in order to avoid the checking correction.

2. Once the dog begins responding reliably to the command "Out, Down" instead of waiting for the assistant to check with the line, then we increase the difficulty of the exercise. The decoy begins stimulating the dog more at the beginning of the exercise, by walking or running away before the dog is sent. The decoy freezes when the "Out, Down" command is given. Now the decoy's behavior begins to provide a "clue" that helps the dog to perform the standoff correctly. When the dog performs the exercise correctly at full speed and in full drive, with the line on and an active decoy, then we are ready to take the line off.

3. Initial off-line standoffs are performed with the decoy standing still. The dog is sent and called "Out, Down." If it performs correctly it is rewarded as before with a "Good" marker and reward in position. If it does not lie down immediately, continuing toward the decoy or even biting, then the handler gives the "No" command (we often have a short, light leash on the dog's collar so that the agitator can reach up and take this leash if the dog bites). Then the handler advances to the dog and delivers a correction, and then we repeat the exercise. It is useful with a dog that is difficult to control to have an "escape hatch" for the decoy (such as a gate or door very, very close by) so that, if the dog disobeys the "Out, Down" and continues toward the decoy, the decoy can quickly escape through the gate and stand passive. In this way we

can prevent the dog from "stealing" reward after disobeying the "Out, Down."

4. Eventually, the stand-off is run with an active decoy, and the dog can be rewarded for a good "Out, Down" by being given the "Yes" as soon as it touches its elbows, allowing it to release forward from the down, and finish its run to the decoy and bite.

GUNSHOTS

A well-trained patrol MWD reacts in a neutral way to gunfire—meaning that it is neither frightened nor aggressive. For both handler safety and mission effectiveness, in almost every case the best thing for the dog to do is to continue doing whatever it was doing before gunfire or explosions and be ready to obey the next command. If, on the other hand, the dog becomes extremely excited and aggressive under gunfire or detonations, this response is highly dangerous for handler and dog. However, this mission requirement is problematic, because the vast majority of MWDs have already been taught prior to procurement to become excited and look for someone/something to bite when they hear gunshots.

- If MWD trainers attempt to teach a dog not to become excited under gunfire by using strong corrections to compel the dog to remain in a sit or down or heel position during gunshots, this procedure can and often does "backfire" in two ways—we can inadvertently teach the dog to be afraid of gunshots (because they become associated with corrections and pain), or we can teach the dog to bite the handler.

- Normally, the best approach to teaching a dog to behave calmly under gunfire is counter-conditioning the gunshots with the ball reward.

  1. In the initial procedure, dog and handler are placed at about 100 yards from an assistant with the weapon. The assistant is protected in some way in case the dog runs to him and attempts to bite. The handler throws the ball across the field and sends the dog in pursuit, and two or three gunshots are fired while the dog chases, but before it picks up the ball. Then the shots are fired after the dog picks the ball up and is returning to the handler. Next, shots are fired while the dog is close to the handler and

moving about with the ball in its mouth. Then shots are fired after the handler has commanded the dog to release the ball, while the dog is waiting for the next throw. In the next stage, the dog is asked to respond to simple commands like sit, down, and heel while occasional shots are fired. Eventually, the dog is required to sit at heel with the ball out of sight in the handler's pocket while a series of shots is fired at 100 yards distance. After each series the dog is rewarded with a throw of the ball.

2. In the next phase, the assistant/gunshots are gradually brought closer. The dog sits at heel and in attention while the shots are fired nearby. If it attempts to break position, it is leash-corrected. When it holds position and maintains attention, it receives a ball reward. When the dog is proficient in this skill, then the handler begins to hold the weapon and dry-fire it while the dog remains at heel. Eventually, the handler heels the dog to the gun lying on the ground, picks the gun up with the dog at attention in heel-sit position, fires a series of shots, puts the gun down, heels away, and then rewards the dog. Eventually the ball reward is provided less and less often, until finally the dog no longer needs any more than praise reward.

3. Trainers must keep in mind that loud gunshots and heavy concussions hurt and cause injury to hearing organs. In addition, dogs appear to be especially sensitive to muzzle blast. Therefore, attempts should be made to protect the dog's ears, and to place them out of muzzle blast and far enough away from the weapons fire to meet local safety standards.

CHAPTER 5

DEFERRED FINAL RESPONSE (DFR)

DFR BACKGROUND

This chapter discusses the DFR method for training substance detector dogs that is now in use in the MWD course at 341 TRS. Because Deferred Final–trained dogs are taught to look at source when indicating odor, rather than at their handlers, this method is frequently referred to as "focus training." This paper is provided to assist MWD users in understanding and troubleshooting DFR dogs.

To become proficient in substance detection, the dog must acquire two main conditioned associations, or lessons. First, the dog must learn to recognize the target odor, meaning that it must begin to expect reward when it smells that odor. Second, the dog must learn that to receive the reward, it has to sit. This sit is, of course, called the "final response" in DoD terminology.

REWARD NOT FROM SOURCE METHOD

The traditional DoD method for teaching odor recognition and final response was called Reward Not From Source (or Reward NFS). The dog is encouraged to investigate an area or a scent box, commanded/helped to sit, and then given a reward (ball/Kong). Over the course of many trials, the dog learns to sniff a series of locations, and sit "on" the location that smells of target odor, without sitting on any locations that do not smell of target odor. Although it appears procedurally simple and therefore practical for students and non-experts, the Reward NFS method in reality relies upon great experience and skill on the part of the trainers. Furthermore, Reward NFS often "builds in" to the dog undesirable behaviors that present long-term challenges to trainers and users. This is because the method:

- Attempts to teach the dog odor recognition and final response simultaneously.
- Makes no attempt to associate reward with odor source.

Two Factors

These two factors interact to bring the dog's attention to its handler during the crucial period when it should be learning to pay attention to odor. The process by which this learning occurs is as follows: During the initial trials, of course, the target odor has no "meaning" for the dog. In Reward NFS the animal is persuaded to more or less accidentally sniff the odor, normally by having the handler use hand presentations to get the dog to place its nose near the training aid. Then, hoping that the dog has "noticed" the odor, the handler manipulates the dog into a sit position, most often with a combination of verbal commands and physical prompts such as leash tugs and/or pressure on the hindquarters. In the course of this intervention, the dog's attention becomes focused on the handler and, once in the sit position, this attention to the handler is rewarded by presentation of the reward.

Handler Cues

Because the dog is rewarded while looking at the handler rather than while sniffing odor, the dog is very slow to learn that odor predicts reward (often requiring more than 150 trials). In addition, it learns that handler behavior also predicts reward. It begins to watch its handler closely during detection problems; it tends to rely upon presentations to make it sniff; and it learns much about the cues (stutter steps, etc.) a handler gives when a training aid is nearby. As a result, the dog often develops a false response tendency based on handler cues, and very much time and effort is expended in trying (often unsuccessfully) to "work out the cues."

Handler-Dependence/Lack of Independent Search Behavior

In addition to slow acquisition of odor recognition and high false response tendency, the most common deficiencies in dogs trained with Reward NFS are handler-dependence/lack of independent search behavior, weak change of behavior in response to odor (the dogs tend to simply sit when encountering odor, making it difficult for the handler to discriminate between false responses and "hits"), and poor localization (many Reward NFS dogs have a "fringing" tendency). All of these weaknesses are consequences of a system that encourages the dog to look to its handler for reward and guidance early in training, before the dog's behavior has come under strong control by target odor.

DEFERRED FINAL RESPONSE METHOD

The DFR method differs from the Reward NFS method in three critical ways:

1. First, during initial training of odor recognition, the final response is not required. Instead we defer teaching of the final response until later, when the dog has mastered some critical pieces of learning. On initial trials, the dog is prompted to search a given area, normally by pretending to hide the reward in that area. When the dog investigates, sniffing for the reward, and sniffs target odor instead, then the reward is provided. Thus we provide the dog with a very "clean" and simple pairing of target odor and reward, without any cues from the handler to compete with odor for the dog's attention. As a result, the dog learns to recognize odor extremely quickly. Dogs trained with DFR normally exhibit strong changes of behavior in response to target odor after 5 or 6 trials.

2. Second, DFR is a "reward-from-source" method, meaning the dog is taught that reward originates from odor source. There are many ways to provide reward from source, including simply placing the reward with the odor, but in DFR we rarely place

the reward with the training aid. Instead, we "lob" the reward in from behind the dog when it is sniffing odor, so that the reward appears without any warning and falls as softly as possible directly on the odor source.

3. Third, during early stages of DFR training, the role of the handler in making presentations and guiding the dog's search is greatly reduced. In fact, the handler gives the dog as little information as possible. The dog searches off-leash in a confined area, or works on leash with an absolute minimum of influence from the handler. We are not concerned if the dog "walks" the training aid. We operate from the principle that the worst thing the handler can do in detector dog training is show the dog where the aid is by stopping the animal on the aid. It is the dog's job to stop the handler on the training aid, and if it fails to do so it learns two important lessons:

- No one will help it to find odor.
- Leaving odor is a mistake to be avoided, because it results in more work and greater delay to reinforcement.

OVERVIEW OF THE DFR TRAINING SEQUENCE

The first step is to teach odor recognition by "paying on sniff" without any requirement for final response. The dog is not paid unless it is clear that it is sniffing and that it has "noticed" the smell of the training aid. Normally within 1 day of training and no more than 10 trials the dog exhibits vigorous changes of behavior when encountering target odor. These changes include bracketing to source, perhaps "freezing" behavior, and sometimes scratching or biting (i.e., "aggression") at the source. Aggression is not desirable and efforts are made to control it, but all recent experience in the Specialized Search Dog and other dog courses indicate that, in a focus-trained dog, a degree of aggressive responding early in training is common and does not necessarily indicate that the animal will develop into a persistent "aggressive responder."

- During the first 6 to 8 days of training and perhaps 30 to 50 trials, the dog works on one odor, with a minimum of handler presentations and interference, with "pay on sniff" rather than final response. Efforts are made to encourage the dog to look/focus/point at odor source, because when the dog focuses on odor it is not observing its handler and learning about handler cues. The dog is desensitized to people in the search area and taught to ignore physical contact and interference from personnel while it is searching. At the same time, it is taught to focus on odor for 2 or 3 seconds at a time and taught to "check back" to odor on command when focus is broken.

- Simultaneously with odor recognition training, the final response (sit) is pre-trained. The sit is not performed in "obedience" mode, and corrections are not used. Instead the dog is induced to sit with food or a reward. Ideally the dog learns to sit when it is pointing at a reward (such as a Kong held by the handler) that it cannot take because it is blocked by a barrier (in this case the handler's hand enclosing the reward). The dog is never forced to sit. Instead it is kept interested in the reward until it chooses to sit voluntarily, and then it is rewarded. Initially, the use of a "Sit" command or any other cue to sit is avoided, so that the sit is not dependent on a command. Ideally, the dog perceives a blocked reward and learns to spontaneously sit to unblock the reward. Once the dog sits quickly and easily to unblock a reward, then we may attach a cue to the sit, such as a light touch on the dog's rump, or a slight tug on the collar, or even the verbal command "Sit."

- When the dog recognizes target odor, stops on it and stares, can be made to "check back," and when the pre-trained sit is fluent, then we begin to require the animal to give the final response on odor. This stage is normally reached on day 6 to 10 of training. Initial final response training can be accomplished in a number of different ways, but in the most straightforward method, the aid is placed in a piece of furniture at about nose height. Rather than paying on sniff, the handler cues the dog into the sit using one of the cues (tug on leash from behind, touch on rump, verbal command "Sit") that was taught during pre-training of the sit. It is important to realize that this use of the word "cue" is not identical to the common use of the word "cue" in DoD terminology. When we speak of a "handler cue" in Reward NFS, we are referring to a piece of information that the handler gives the dog to help it sit on/find the training aid (normally the "handler cue" is a mistake or an accident). In DFR we never deliberately "cue" the dog to help it find the training aid. The dog must always find the training aid and stop on it without assistance. The "cue" in DFR is merely help from the handler in completing the final response. One way to think of it is if the dog independently detects and localizes the training aid, then it "earns" help with the final response.

- The dog is introduced to additional odors, beginning with pay on sniff. Once the dog begins to show recognition of the new odor (normally within three to five trials or even less), then we add the final response to the new odor. The dog also is taught to perform search exercises in new training areas (vehicles, aircraft, etc.). Finally, as one of the last important lessons the dog acquires, the handler for the first time encourages the dog to look to him for help in finding the training aid (i.e., the dog is trained to accept guidance in the form of handler presentations of productive areas).

LIABILITIES OF DFR

Above we reviewed the significant liabilities of the standard DoD (Reward NFS) method. DFR also has liabilities—it can produce certain undesirable behaviors. However, these undesirable behaviors are not as harmful and difficult to correct as the handler-dependence, weak change of behavior and localization, and high false response tendency so common in Reward NFS dogs. *In a word, the kinds of problems associated with Deferred NFS are the kinds of problems we would rather have.*

- DFR-trained dogs can be excessively independent while searching, and may sometimes be reluctant to accept presentations. This shortcoming is offset by the fact that these dogs are often extremely accurate and effective when "scanning" and working independently.

- The most common problem in DFR dogs is a tendency to stop/stand and stare at odor rather than sit. It is worth noting that this is not a functional problem from the standpoint of detecting narcotics or explosives. A distinct change of behavior followed by localization and then a stop and stare is every bit as recognizable to the handler as a sit. In fact, the DoD Specialized Search Dog Program officially authorizes the stop and stare as a final response.

- The first thing to realize about "stop and stare" is that it is the result of physical and psychological tension rather than disobedience. For the DFR dog, an odor source is like a magnet that draws the dog's attention. The dog orients to odor in the same way it looks at the reward—with great excitement. As a result the dog carries much more physical tension in its body than a Reward NFS dog. This tension makes the final response more difficult because, to sit, the dog must relax major muscle groups in its back and hindquarters.

- The expression "stop and stare" is used to describe a wide range of difficulties with the final response, ranging from a dog that delays its sit for 5 or 6 seconds, to an animal that exhibits a classic "locked up" stop and stare, in which its body goes completely rigid. The classic "locked up" stop and stare is a comparatively rare problem with DFR dogs. More common final response issues are:

1. Slow sit.
2. Not completely reliable sit (stops/stands and stares sometimes).
3. Refusal to sit unless cued somehow.

SLOW OR RELUCTANT FINAL RESPONSE IN DFR DOGS

- If the dog's sit is slow or delayed, but all that is needed is to wait a few seconds for the final, we advise that you leave well enough alone. There is no compelling reason to pressure the dog for a rapid or "crisp" sit; so long as the animal stops on odor without any help from the handler, will not leave source, and sits on its own without assistance given a few seconds.

- If the dog sometimes does not complete the final response (this may happen when the footing is difficult, or the aid is very low to the ground, or the dog is tired, etc.), then it is important not to overreact. A stop and stare error is not equivalent to critical errors like missing an aid or false responding. Keep in mind that the dog has completed the most important part of the job: It has found and indicated the drug or explosive hide. Now we just need the animal to fulfill the statutory requirement for the final response. Strong verbal corrections ("No!") and physical corrections (jerk on choke collar) are normally not helpful because they increase the dog's stress and tension, and make it all the more difficult for the animal to relax enough to finish the final response. Likewise simply trying to force the dog's hindquarters down to the ground should be avoided, because it generates resistance and makes the dog lock its legs.

- The first corrective action for a dog that stops and stares should be familiar, because it is a fundamental part of handling any DoD-trained detector dog. When the dog stops on the training aid, the handler should keep moving away from the dog and present the next location in the search area. If the dog neither leaves the training aid nor sits, then the handler should put more pressure on the dog with slight leash tension or a tug on the leash. If the dog is not actively working the odor, or sitting, then it needs to move on and actively sniff the next location. This procedure is no different than handling a Reward NFS dog, but still many handlers need to be reminded of it. In this situation, many dogs will react by completing the final because they do not want to be taken away from the aid. Some dogs will allow themselves to be taken away from odor but then they immediately realize this was a mistake. They begin trying to cut back to it, and if they are allowed to do so (on their own—the handler does not "bounce them back" to the aid), they will complete the final response when they get to source.
- If the above procedure is not effective, and still the dog stops and stares, then the next thing for the handler to try is to simply give a "Sit!" command from behind the dog. Very many animals will comply. In preparation for this procedure, it is a good idea to practice and "fine tune" the dog's response to the "Sit!" command during obedience training.
- If the dog does not complete the final response with the verbal command, then the handler can try two other types of assist. First, he should use the leash to administer light snaps or tugs on the choke collar, straight back toward the dog's hindquarters. If this is not effective, then the handler should try lightly tapping or pushing downward against the dog's rump, just ahead of its tail and behind the hips. The handler should NOT push against the dog's back ahead of the hips, because this normally causes the dog to brace against the pressure. Most dogs know one of these assists for the final, or a combination of them may be most effective, and will respond by executing the sit. Once the dog has completed the final response, then the handler praises with "Good!" and prepares to provide the reward.
- The reward should be given while the dog is focused on odor, from straight behind or some other blind spot, and it should be lobbed in as softly as possible. Ideally, it goes "dead" at odor source, so that the dog can

simply pick it up, rather than chase it frantically around the room. Specialized search dog (SSD) trainers and other specialists in DFR sometimes arrange nets or even pillows near the training aid, or use a "spiked" tennis ball or something similar, so that the reward lands at source and stays there without bouncing. If the assist distracted the dog, then focus on the odor source should be reestablished prior to reward.

- If the sit is consistently assisted as described above, the dog's final response normally improves and no other special efforts are required. Sometimes it may happen, however, that the dog becomes dependent upon the assist—it will not sit unless it is given some cue. In this case, it is often effective to allow the dog to make repeated finds of the same training aid. This procedure removes the search element from the exercise and allows us to concentrate the dog's energy on the final response. Also, because the dog is asked to sit repeatedly in the same location, the animal's anticipation works in our favor to produce a "fluent" (relaxed, easy, prompt) sit. The first time the dog finds the aid, it is assisted into the sit and paid. After the reward is recovered from the dog, the animal is allowed to go right back to the aid. On the second indication of the aid, if the dog does not complete the final response, then it is assisted, but the dog is not paid. Instead it is praised for the sit, then gently pulled backward off of the training aid. (Keep in mind here that a well-trained "focus" dog will not allow itself to be easily taken away from a training aid. Instead it will fight to "check back" to the aid.) Once the dog is pulled away, it is encouraged by praising it, and then allowed to go to the aid again. If the cue is still necessary to produce the sit, then the dog is again praised for the sit verbally and physically, then pulled gently away from the aid, and allowed to go back again. The procedure is repeated as many times as necessary (within reason). When the dog eventually sits without the assist, the handler praises it and pays. Once they understand that they must sit without the assist to be paid, most dogs quickly improve their final response.

- A word of caution is necessary here. Unless the dog is well-trained in "checking back," pulling the animal off of the aid repeatedly can produce some undesirable side effects. Some dogs may become discouraged or frustrated, and begin to leave the odor source. If a cage is used (after a bit of basic training to establish good focus on the visible aid), then the dog is less likely to leave the visible aid. However, the dog may not leave the aid, it may instead become more tense and resistant, and begin stopping and staring more. However, the resistance may be specific to the aid that it has encountered again and again. If we move the aid to another location and "surprise" the dog with it, the animal may show one of its best unassisted sits. Similarly, if we have a session one day in which we "hammer" the sit with many repetitions, even though the dog may conclude the session looking as though it has regressed, sometimes we find that on the first aid of the next training session, the dog will give a crisp, perfect final.

STOP AND STARE IN DFR DOGS

If the dog's sit is extremely slow, or unreliable, or if the dog has become very tense, "locking up" and going rigid on the training aid, then we must take a more patient and multipronged approach to improving the final response. Probably the most important part of this approach is making sure that the dog knows how to sit to gain access to a blocked reward. This process is referred to as "pre-training" the sit. It was mentioned briefly above, but now we will discuss it in detail.

Pre-Training the Sit

1. First, the dog is taught to sit when the handler holds a reward in the air above its head. If possible, it is better to avoid use of the verbal cue "Sit." Instead just keep the dog's attention and let it experiment with various behaviors until it hits the right one. When the dog sits, the handler gives the reward, preferably by simply handing it to the dog calmly or dropping it into the animal's mouth. The reward should NOT be vigorously thrown or bounced. With many dogs it is more constructive to begin pre-training the sit with food rather than the ball/Kong reward, because food produces a more manageable level of drive. In addition, in the process of taking and eating food, the dog can more easily be taught to "respect" (i.e., not bite) its handler's hands.

2. In the next step, the reward is held lower, where the dog can reach it, but the handler "blocks" the reward by enclosing it in his hands. The dog can see the reward, smell it, even get nose and teeth on it, but not take it. We use the hands to block the reward rather than a physical barrier because most dogs respect their

handler's hands—they do not "aggress" a reward that is held in the hands the way they would a reward held in a drawer. In this situation, with the reward just inches away, nearly every dog falls into stop and stare behavior very similar to the behavior we see in detection. The handler does not try to hurry the dog into the sit. Instead he provides time for the dog to "think," just keeping the dog's attention on the reward by moving it slightly when necessary. When the dog eventually sits, it receives the reward (i.e., is "paid").

3. As time passes, the handler holds the reward lower and lower, trapping it against the legs or body, waiting until the dog sits and then paying the animal calmly. Once the dog is very "fluent" with this behavior, easily and quickly offering a relaxed sit to unblock the reward, then the handler begins to hold the reward against the wall or against furniture, teaching the dog to give the sit whether the reward is held six feet up the wall or on the ground. Low placements of the reward, like low placements of a training aid, are particularly difficult for the dog. Note: Blocking the reward by holding it against walls and furniture should be performed sparingly because the dog can confuse this gesture with hand presentations, causing a tendency to sit or "false respond" on hand presentations that will then have to be extinguished.

4. Next the handler gives the reward to another party. The handler holds the dog on leash while the assistant shows the dog the blocked reward, and pays the dog for the sit. The assistant begins with the reward held high, but then quickly passes through all the steps of holding it trapped against the body and then legs, and then against a wall or furniture at various heights.

5. Eventually, the reward is not held by a person. Instead it is "trapped"—jammed between a piece of furniture and the wall, or between a car door and its frame, etc. The exact situation does not matter, only that the dog can see and smell the reward but not touch it or take it. The moment the dog sits, then the door is pulled open or the furniture pulled away from the wall to liberate the reward. If the dog responds aggressively rather than sitting, then you wait until the dog becomes discouraged, stops scratching/biting, and sits, and then you pay. The most important factor

is not whether the dog is allowed to "aggress" a reward or aid—it is whether the dog achieves some result with the aggression. With the vast majority of dogs, if aggression does not produce any result such as opening a drawer or moving furniture or making the handler react, then alternative behavior (i.e., sitting) will quickly take over if it is rewarded. Occasionally, it may be necessary to pay the sit not by liberating the trapped reward, but instead by lobbing another reward in from behind the dog.

6. As a last step, the fluent and relaxed sit is placed under the control of some cue. This cue can be a voice command, a tug on the leash, or a touch on the hindquarters, whatever works best with that dog—but the cue must be given from behind the dog while it is facing a blocked reward. The exercise is set up as before, with a reward trapped in a suitable location. The dog watches the reward placed, and then is allowed to move forward on leash until it is stopped by the barrier blocking the reward. The handler gives the cue as the dog is assuming the sit position. This means that the cue does not cause the sit—the sit is voluntary. But by pairing the cue with the voluntary sit, you can gradually give the cue the ability to trigger the sit. Most important, the sit we obtain is the relaxed, fluent sit the dog has learned in pre-training, rather than a tense, resistant sit.

Reteaching the Final Response

1. Once the sit is adequately pre-trained, with a cue that enables the handler to sit the dog from behind when it is facing a blocked reward, then we bring odor into the situation. We choose a location in which the dog has been extensively drilled on sitting to unblock a reward—a "trap" between a cabinet and a wall, or between a car door and its frame, etc. We place both the reward AND a training aid in the trap. We must be careful to place the reward and training aid at a height that facilitates the sit—normally at about nose height.

2. The dog associates the particular location of the trap with sitting. It also associates a blocked reward with sitting. These conditioned associations help to overcome the dog's tendency to respond to odor by freezing and staring.

3. The dog is allowed to see the reward and the training aid being placed in the trap. When the dog approaches and sniffs at the reward, it also catches target odor. Depending on the individual dog and the preferences of the trainers, we can then do one of two things:

   a. Wait until the dog sits. We wait as long as necessary without giving the dog any input or cue. We watch for the sit and ignore anything else that the dog does, except that, when the dog becomes so frustrated it leaves odor, the handler may entice it back to the trap again. This process may take a very long time (up to 5 or 10 minutes) and it may get "ugly," in the sense that the dog may stand and stare for minute after minute, or begin looking around, even leave odor and return, or it may begin scratching or biting at the trap. However, nothing produces so much learning as a problem independently solved, and if we allow the dog to choose the sit in its own time and in its own way, just a few trials can result in dramatically less stopping and staring and a faster and faster final response.

   b. Once the dog sits, then it is rewarded—not by unblocking the trapped reward but instead by lobbing another reward in from behind. After a bit of reward-play, the dog is immediately sent back to the exact same trap to practice the final response again. In fact, the dog is sent back to this same trap several times in a row. This is not a search exercise, it is a final response exercise, and knowing the location of the aid only helps the dog.

4. For the second session the exercise can be moved to another location, so that the dog has to search a little before finding the trap. This will revive the animal's sniffing behavior so that on the first trial or two we can be sure that it is smelling the training aid (and reward), rather than using eyes alone.

5. Cue the sit. If the dog's sit (to unblock a reward) is controlled by a cue, then the trainers can also cue the sit rather than wait for it. This method has the advantage that it results in a faster sit and less "ugly" behavior like leaving odor or aggressing, but it has the disadvantage that the cue will eventually have to be

extinguished, which may involve a surprising amount of repetition and work.

6. As before, the dog is sent to a trap containing a reward and a training aid. The moment the dog sniffs, it is cued by the handler into the sit. The handler does not wait to see if the dog will stop and stare; he immediately cues the sit without any pause, almost hurrying the dog into the sit. Also, once the handler begins to "ask" for the sit with the cue, he makes sure that the sit is obtained, even if in the process the dog loses focus on the training aid. When the dog completes the sit, a reward is thrown in so that it lands at the odor source (training aid odor and trapped reward odor).

7. This exercise is repeated many times, sometimes leaving the trap in a familiar location so that associations with that location help the dog to sit, but moving it often enough so that sniffing behavior is from time to time reestablished. Every now and again, the handler tests the dog by waiting a moment to see if the dog begins to sit before being cued. Then for the next few trials, the handler returns to immediately cuing the sit again without waiting to see what the dog will do on its own. Once we find that during the test trials the dog has begun to reliably initiate the sit without waiting for the cue, then we move to the next step.

8. Throughout the trials described above, the dog was paid once it reached the sit position whether a cue was necessary or not. Now we raise the criterion—if a cue is required to obtain the sit, we will not pay the dog. The dog is allowed to find the reward/training aid, and the handler watches to see if the dog gives the final. If not, the handler cues the sit, then praises the dog verbally and perhaps physically while in the sit, then pulls the dog by the leash away from the aid 5 to 10 feet, and then allows the dog to go back to the trap. This procedure is repeated again and again, cuing the sit if necessary but never paying the dog for the sit if a cue was used. When the dog finally sits on its own without the cue, we give the reward.

9. Eventually, once the dog is very fluent with the sit when presented with the reward and training aid together in a trap, then

we omit the reward and present the training aid alone. We follow the rules described in the last paragraph, giving the dog only praise and encouragement if it requires the cue in order to sit, but giving the reward if the dog sits without a cue.

Waiting the Dog Out versus Cuing the Sit

1. Choice of strategy is guided by trainer inclinations and dog characteristics. If the dog has a very rigid, "locked up" stop and stare, but will not leave the aid and does not tend to aggress, then the waiting approach is often best. Attempting to cue such a dog often stimulates more tension and resistance. Instead the animal just has to be given the time to learn that stop and stare does not produce reward, but sitting does. It is not unknown to wait 10 or 15 minutes.

2. If the dog has a very prompt, fluent, unresisting sit when cued, and if the dog is not extremely tense when indicating a training aid, then the cued approach is likely to be quick and effective and has the additional advantage that it will prevent the dog from aggressing the aid, leaving it, looking around, etc.

Use of a Cage to Contain the Training Aid

1. Rather than the trap, DoD's Specialized Search Dog Course often places the training aid inside a sturdy "cage" of some sort that allows the dog to see the training aid. Although being able to see the aid may decrease sniffing somewhat, visual access to the aid helps to establish and maintain focus. This is helpful because maintaining focus can be very difficult while retraining the final response, because we may leave the dog "on the aid" for long periods of time while we wait for the sit, or subject the dog to strong physical influences to cue the sit.

2. The addition of visual information to the situation can have a beneficial effect on many dogs that have difficult habits when working "hidden" training aids. Dogs that are very aggressive to aids in other situations fall into staring behavior when they are caged; dogs that walk away from hidden aids in frustration stay with them when they are visible inside a cage.

3. Initially the cage is placed on the ground in an empty corner. A few introductory cage trials are performed to establish the dog's focus in this unique situation. The dog is paid "on sniff," which may seem puzzling in view of the fact that the goal is to cure a stop and stare problem. However, a few "pay on sniff" trials will help us establish focus on the caged aid, and they won't make the stop and stare problem any worse than it already is.

4. An assistant shows the dog the reward and pretends to place the reward on/in the cage. The dog is released from a distance of 15 or 20 feet away. As the dog moves in to investigate the cage the assistant "fades away" slightly so that, just as the dog reaches the cage and sniffs, he can drop the reward in on top of the cage without the dog seeing where the reward came from.

5. After two or three trials like this, the dog's sense of sight will assert itself. The animal will begin sniffing less and looking more. It may do any number of things. The dog may stop short of the cage and stare at it, or it may go to the cage and check it and then turn away or begin to look around. In response, the assistant encourages the dog to investigate the cage, often by presenting it with his hand, and then drops the ball in on top of the cage when the dog "checks" it. The most important consideration is to pay the animal when it is focusing on the caged aid.

6. After several trials, when the dog's sniffing has been reduced because it has begun to look at the training aid/cage instead of smelling it, then you can "wake the dog's nose up" again by moving the cage to a new location. You can move it to another corner of the room—or you can place it inside a chest of drawers, replacing one of the drawers with the cage. In this location, the cage and training aid will be easy to find, but the dog will still have to sniff to find it. In addition, once found, there are plenty of visual cues to help the dog stay with the training aid. The dog will go to the familiar corner, find the cage is gone, and begin sniffing for it. When it finds the cage it will check the aid closely and sniff, and you can pay for this sniffing and checking. After a few trials like this, you should obtain good "focusing" and "checking back" and sniffing behavior on the cage. Now, if you withhold the reward, you should see the dog's

stop and stare behavior in full force. At this point you can adopt one of two strategies:

a. If you choose to wait for the sit, then the visual cues of the training aid inside the cage will help to keep the dog with the aid while it "thinks" the problem over. In addition, the cage will protect the aid and prevent the dog from getting any result should it become frustrated and begin scratching and/or biting at the aid.

b. If you choose to cue the sit, then again the cage will help keep the dog focused on the training aid while you use the cue. This is especially important if the dog loses focus while you are cuing the sit. Once in the sit, with a little encouragement the dog can be induced to look back at the training aid and then be paid for checking it. You assume you will have to assist the dog many times. Initially you pay the dog even if you have to assist it with the cue. Later you pay only if the dog completes the final response without assistance.

ISSUES WITH DFR DOGS

Disrupted Focus

To understand the importance of maintaining focus in the DFR dog, you must again consider the common shortcomings of the Reward NFS dog. Two of the cardinal faults of a conventionally trained Reward NFS dog are that the animal:

1. Tends to sit on fringe odor rather than going to source.
2. Tends to stop working odor/sniffing after the final response.

Two Faults

These two faults, both rooted in the fact that the dog expects reward from the handler rather than from source, interact in the following way to produce final responding on handler cues rather than on odor. When the dog smells odor, it final responds and looks at the handler. If the dog is not as close to source as you desire, then the handler asks the dog to leave the final and continue to sniff/search. The dog normally does not do so effectively, because once it has smelled odor it quite rightly expects reward to come from the handler, and so it looks at the handler, its mind on visual cues rather than sniffing. The handler normally makes a hand presentation or gives some other sort of encouragement; the dog makes a token head movement while looking at the handler, moves in the direction the handler has gestured, and then immediately sits again. If the dog has moved closer to source, the dog is then paid. However, from the moment the dog first gave final, it has not done any further sniffing or processed any more odor. It has merely oriented at and responded to a series of handler cues, and then been rewarded.

DFR Dog Needs No Encouragement

In contrast, a DFR dog needs absolutely no encouragement to go directly to source. In addition, if the dog is well-trained, it continues to work odor and investigate source even after first final. This is the basis of the "checking back" behavior that is so valuable to the detector dog handler. When the dog "checks back," it points at, or crowds, or drives into the odor source, normally in response to some attempt on the handler's part to take it off of odor. This checking back behavior is extremely useful because it allows us to:

1. Refocus the dog's attention on odor after it has been disrupted by, for instance, cuing the final response. This enables the handler to pay the dog for sniffing odor rather than looking at its handler.
2. Ask the dog to "confirm" a find. You ask the dog to confirm by trying to pull it away from something it is interested in, either before final or after final. A well-trained DFR dog must be physically dragged away from odor and will exert strenuous efforts to get back to odor and repeat the final. This makes the dog extremely easy to read, and simplifies the problem of telling the difference between interest in a distracting odor and a change of behavior in response to a bomb or drug hide.

Lost Focus

When a DFR dog has lost focus on training aids, when it no longer checks back when asked, this is normally as a result of "damage done" while teaching the final response. However, note that many dogs lose focus while learning the final and have to be "rebuilt" once the final is fluent. This is not difficult as long as the dog was taught to check back to odor prior to introduction of the final response.

1. The first thing to do in reestablishing focus is to run a few "pay on sniff" trials. It is very helpful to have a visible training aid (in a cage) so that the dog has a visual fixation point. Next, allow the dog to find the aid again but wait for the final response. Once the dog has completed final, then the handler can attempt a variety of different techniques to induce the dog to "check back."

2. Stand behind the dog and wait for a few seconds, encouraging the dog verbally to "Check!" Be extremely alert for a head movement toward the training aid, and be ready to pay on odor source.

3. Exert a slight backward pull on the dog's collar, as though to pull the dog straight back off of odor. At the moment the dog feels this traction, it is likely to "head-poke" back at the aid, and at this instant the handler or an assistant must pay by dropping the reward on source.

4. If slight pressure does not cause the dog to check back to source, then the handler can pull harder and actually pull the dog backward out of the sit. Praise and encourage and excite the dog and then release it to go back to source. Initially, you pay the dog for checking back to odor without demanding the sit. Later you withhold the reward until the dog checks back and then completes the second final response. (Note: The decision about whether it is "safe" to reward the dog on sniff without the final response depends on how fluent the dog's final response is—if the dog has a very fluent, relaxed sit, then it is normally not harmful to sometimes pay on sniff. If, on the other hand, the dog has a resistant, tense sit, then paying on sniff may quickly result in stop and stare.) If the dog completes the second final but loses focus again, then give the slight pull (as above) to try and reestablish focus during the final. However, it may be impossible to obtain both focus and final at the same time, so alternately reward one and then the other until the dog unites them both.

5. If the dog has completely lost focus on the aid so that none of the above produces the check-back, then the handler can attempt

a hand presentation to bring the dog's attention back to the aid. Do not pay the dog while the hand is still on source. This merely rewards the dog for looking at a part of the handler's body. Instead tap on source to draw the dog's attention, and then attempt to "fade" the hand out and pay for focus on source.

6. One trick used by the SSD Course is to place a radio with the training aid. When the dog is in final, the radio is keyed. The dog looks at source and is paid.

7. Another trick to reestablish focus is to use a "reward system" to deliver the reward from odor source. There are very many such devices—some complicated and elaborate and some very simple. In one version used by the SSD Course, a training aid is placed in a box that opens toward the dog. Poised two or three feet above the aid is a reward in a PVC pipe, held there by a piece of thin bungee cord stretched from one side of the tube to the other. The reward rests on top of the bungee. Another piece of bungee cord or string is tied to the middle of the first so that if someone pulls sideways on it, the reward can slip past the first bungee and down the tube and land on the odor source. The important elements of such a reward system are that the reward is separate from odor prior to payment; that the dog is paid by someone other than the handler actuating the reward system; and that the reward lands as directly and softly as possible on the odor source.

Aid Purpose

As a final note, keep in mind the purposes of focus on the aid: to make sure that the dog does not learn "handler cues" by watching the handler, and to make sure the dog can be rewarded *on odor source for approaching source*. If both of these conditions are met, then you are likely to produce a dog that shows strong independent search behavior, an obvious change of behavior, good bracketing to source before final, and a desire to check back to odor if walked or pulled off of it. If you have these elements in place, then exactly where the dog focuses while in the final response is not especially important as long as the animal does not look directly at the handler during payment. In fact, many DFR dogs learn a sort of "superstitious staring" behavior in which they do not focus on source (especially when it is high overhead and well-hidden); instead they stare generally upward or away from the handler. This behavior still serves the most important purpose of focus on odor source—it keeps the dog from being paid for focusing on the handler.

Refusal to Accept Presentations

1. Because early in training they were never encouraged to look to their handlers for any help or information, DFR dogs tend to be very independent while searching. Many of them work very effectively on loose-leash "scans," and handlers find that this is often the best way to begin a search. They make presentations only when the dog misses a productive area. Eventually, if the dog is unable to make the find independently, or when it becomes tired and needs some guidance and encouragement, then the handler becomes more active and supplies more guidance.

2. If a DFR dog is excessively independent and will not accept its handler's presentations, this is normally because the dog has simply not yet learned that the handler can be a help in finding the aid. (Note that this is to some extent deliberate—you do not want the dog to learn to "use" presentations by its handler until late in training, after it has already become completely independent, aid-focused, and self-reliant.) In working with such a dog, the two most important factors are: The handler should give few presentations, and the handler must make sure that these presentations are productive from the dog's standpoint.

3. Set up a search problem with one training aid. Begin the search with the dog working independently, without presentations, but stay away from the area where the training aid is hidden. Wait until the dog tires substantially, and becomes more receptive to the handler's influence. Then move the dog into the area of the training aid and make a careful presentation in a productive area near the training aid, so that the dog catches the odor and completes the find and receives a reward. Repeat this sort of search exercise a few times—run the dog on a long problem without any possibility of finding the training aid until the dog wears down somewhat, then take the dog into the vicinity of the training aid and make a presentation or two that put the dog "into odor." The animal will quickly learn to look for presentations, because

they indicate an area in which odor can be found. Once the dog becomes very receptive to that initial presentation, then the handler can begin carefully adding a few more presentations here and there, very gradually building up the number of presentations the dog will eagerly accept before finding the training aid.

4. When teaching the dog to accept high presentations, it is especially important to make sure that the first few presentations are productive for the dog—that they help the animal find odor. Otherwise the dog quickly learns to merely rear up against anything that is presented, but without sniffing.

CHAPTER 6

DETECTOR DOG TRAINING VALIDATION AND LEGAL CONSIDERATIONS

VALIDATION TESTING

The kennel master conducts validation tests on each detector dog team annually (not to exceed 180 days from last certification) and prior to initial team certification. These tests are intended to verify the detection accuracy rates annotated on training and utilization forms. Conduct validation testing in a non-task-related environment. Kennel masters and commanders will ensure handlers are provided sufficient dedicated time to complete all validation trials. Make every effort to complete validation within 5 duty days. If a team fails to meet the minimum accuracy rate, the kennel master will immediately initiate remedial training. Retest previously identified remedial teams in unsatisfactory areas only when deemed appropriate by the kennel master. If the team fails upon retest, consider those actions outlined in AFI 31-202/Military Working Dog Program.

- Because validation testing is intended to verify accuracy of the entries on training and utilization forms, the rating for each training area and odor will reflect GO or NO GO. The minimum standard, however, will be an overall accuracy rate of 90 percent for drug dogs, and 95 percent accuracy rating for explosive dogs. (**NOTE:** Percentage rates are based on total number of training aids planted/found, and not the individual percentage rate of each substance/odor planted/found. Should the team achieve the overall standard, but demonstrates difficulty in accurately detecting a particular substance/odor, proficiency training must increase for this particular odor.) All available odors must be used during validation testing; any training aid/odor not used during validation testing period must be reflected in the validation report with a general statement explaining why they were not used. Forward this report to the chief security forces (CSF) who will endorse it and return it to the kennel master. File the report with other probable cause documents and document validation testing on AF Form 323 (see data on form later in document).

- Conduct at least two trials per odor and one negative test (no training aids planted) per validation for both drug and explosive dogs. Drug dogs are also required to conduct 2 residual odor tests. Validation testing must be conducted in at least three of the below areas, but efforts should be made to conduct testing in as many areas as possible:

  1. Vehicles.
  2. Aircraft.
  3. Luggage.
  4. Warehouse.
  5. Buildings/Dormitories.
  6. Open area.

- Validation reports must include date of each validation trial, start time of each validation trial, location of each validation trial, type of area used for each validation trial, aids planted/found for each validation trial, total search time of each validation trial. List each validation trial separately in the report. Validation testing will be documented on the AF Form 323 (see data on form later in document). Document validation testing on the AF Form 323 in the same manner you would for your normal detection proficiency training. Ensure annotation reflects, "Validations conducted by (Grade/Name)."

- Conduct out-of-cycle validation trials if there is any reason to suspect a dog's detection capability has significantly diminished. Out-of-cycle validation and certification trials will also be conducted whenever a detector dog has not received detection training with 1.1 munitions or drug training aids for 30 or more consecutive calendar days.

For clarification purposes of this manual, the term *deployment* is considered when a team is operationally forward deployed in support of contingency or humanitarian operations. When a home station certified team deploys for more than 179 days and returns to their home station within their current home station certification window, conduct out-of-cycle validation trials within 20 duty days of team being reassigned to home station operational detector dog duties. If required accuracy percentages rates are achieved, the team retains its current home station certification status. Should validation standards not be achieved, the team's current home station certification status is revoked and the kennel master will reinstitute a new validation and certification process, thus establishing a new validation/certification cycle for the team. Teams deploying for less than 179 days are not required to revalidate unless, while the team was deployed they did not receive detection training using 1.1 explosive training aids or narcotic training aids for 30 or more consecutive calendar days. Deploying handlers are required to document their animal's utilization and training on AF Form 321 and AF Form 323 while TDY/deployed. Handlers will turn in their TDY/deployed AF Form 321 and AF Form 323 (see data on forms later in document) to the kennel master upon return.

LEGAL ASPECTS

Prior to using the response of a detector dog as probable cause to grant search authority, the team will demonstrate their ability to detect the presence of all substances (odors) the dog is trained to detect. The individual(s) having search granting authority over the installation is encouraged to witness this demonstration. The search granting authority may delegate his responsibility to witness this demonstration to the CSF. After being satisfied with the team's detection capabilities, the CSF prepares and forwards a memo to the search granting authority describing the conduct of the demonstration to include the odors used and the accuracy rate of the team. The CSF will include a copy of the most current validation report, training and utilization records (covering the period since the last validation), and a resume of training and experience for each team being considered for certification. If the search granting authority concurs with the findings and recommendations of the CSF, he will endorse the memo indicating concurrence and return it to the CSF. File this memo with the most current validation report in the dog team's probable cause folder. A similar memo will be prepared for search granting authority signature if he personally witnesses the demonstration and filed in the probable cause folder as well.

- Installation commanders and military magistrates appointed in accordance with AFI 51-201, paragraph 3.1 (see data on form later in document) are encouraged to periodically assess the reliability of detector dogs in a controlled test environment. Such assessments will bolster any probable cause search authorization based on an alert by the dog(s) observed. These assessments are not required and the failure to perform them will not, by itself, invalidate a search authorization made by these officials.
- Conduct a recertification demonstration annually or whenever a handler change occurs.
- SF supervisors and commanders responsible for managing and employing MWD assets must understand home station certifications are only valid under the purview of the search granting authority for that particular installation. For example: A dog team assigned to Lackland AFB, TX, is certified with the search granting authority for Lackland AFB. This certification is only valid at Lackland AFB. Should this team go TDY to Randolph AFB, TX, to perform probable cause searches, regardless of the TDY length, the team must certify under the search granting authority for Randolph AFB prior to conducting searches.

  1. There may be instances such as mission time constraints, non-availability of training aids at the TDY or deployed location, etc., that may prevent an actual certification demonstration to take place. Should this be the case, ensure a copy of the team's probable cause folder is sent with the team along with an official memorandum with the search granting authority's signature block (TDY/deployed location). This memorandum must indicate the TDY/deployed search granting authority has reviewed the team's probable cause folder and finds the documented performance and training of the dog team satisfactory, and the team is authorized to perform probable cause searches within their jurisdiction. The TDY/deployed search granting authority will retain this memorandum.

2. SF MWD teams **must at a minimum** meet validation standard at home station prior to going TDY or being deployed where it is anticipated the team will perform detector dog duties. Once at the TDY or deployed location, it is the responsibility of the gaining unit to conduct the certification process with the gaining search granting authority if required. If so, home station validation trials can be used in lieu of TDY or deployed location validation testing.

3. USAF MWD support to sister services. In today's ever increasing inter-service cooperation and support, it is not uncommon for SF MWD assets to be called upon to assist sister services' and vice versa. Should this occur, MWD teams will validate and certify (if necessary) under the Air Force standard and not the validation and certification standard of the supporting service while being operationally assigned to another service. U.S. Army, U.S. Navy, and U.S. Marine Corps dog teams supporting USAF operations using an USAF search granting authority will validate and certify under their respective service requirements. In most cases, these sister service teams will have already met their home service MWD certification standards prior to departing their home station. Should a sister service dog team be required to conduct probable cause searches under the purview of an USAF search granting authority, follow the guidance outline in no. 1 paragraph above of this manual. All DoD MWD teams maintain certification or probable cause folders. Use the USAF MWD probable cause folder as guidance, but not as an all-inclusive baseline when reviewing sister service dog team documentation.

CHAPTER 7

THE MILITARY WORKING DOG PROGRAM

DOCTRINE

The MWD is a highly specialized piece of equipment that supplements and enhances the capabilities of security forces personnel. It is a unique force multiplier and provides security forces another level on the "use of force" continuum.

FUNCTIONAL AREA RESPONSIBILITIES

Every level of command must ensure the MWD program is efficiently managed and develop expertise to properly employ MWDs. If not properly maintained, MWDs lose their skills rapidly; therefore any planning for long-term use of MWDs must always have the training of the dog teams kept in mind. When employed as an integral part of the security forces team, the entire security forces effort is enhanced.

Installation Commander

AF wing commanders are normally an AF installation's search granting authority because they exercise overall responsibility and control of an installation's resources and its personnel. Wing commanders may delegate their search granting authority to lower echelon commanders such as the mission support group commander. Consult with your installation's staff judges advocate (SJA) for clarification of your particular installation's search granting authority to ensure all legal parameters associated with the MWD detection certification process are met.

Chief of Security Forces (CSF)

The CSF ensures MWDs are properly employed. The CSF establishes guidelines to ensure MWDs are properly trained and integrated into the unit's mission.

Flight Chief

The flight chief ensures MWD assets are properly employed, working hand-in-hand with the kennel master to maintain the team's proficiency at optimal levels. The flight chief will ensure adequate time is provided for assigned handlers to accomplish daily required activities including but not limited to dog team proficiency training, kennel care, and annotating MWD records when providing supervision and management of their flight operations.

EMPLOYMENT AREAS

The MWD team is a versatile asset to a security forces unit and can be effectively employed in almost every aspect of a unit's security, air provost, and contingency operations. Local unit operating instructions address the use of dog teams. Consider them in the following areas:

Nuclear Security Operations

The MWD can be an invaluable asset in the protection of nuclear weapons and critical components. An MWD team may be used in weapon storage areas to replace or augment sensor systems, as a screening force in support of aircraft parking areas, or in support of convoy and up/down load operations. Explosive detector dog teams are highly effective in searching and clearing nuclear operation work and support areas and related equipment.

Air Provost

MWDs detect, locate, bite/hold, and guard suspects on command during patrol activities. They assist in crowd control and confrontation management and search for suspects both indoors and outdoors. Dog teams should not be used as the initial responding patrol for drunk driving traffic stops or dispatches, domestic violence responses, or disturbance responses, if at all possible. These types of responses require SF members to have direct contact with either the subject/suspect(s). Because of this close proximity to the suspect/subject during the initial response phase, handlers would be left with the choice of either focusing their full attention on the subject/suspect or their dog. Should the handler choose to focus their attention on the subject/suspect, rather than the dog, this could result in a lack of effective control over the MWD with the MWD inadvertently biting the subject/suspect or put the handler at risk should the subject/suspect become violent. In most potentially dangerous law enforcement responses, dog teams are well suited to provide backup or as a secondary response patrol. In many cases, just the mere presence of a dog team within the immediate area as an over watch will deter most hostile or violent acts.

Drug Suppression
MWD teams specially trained in drug detection support the Air Force goal of drug-free work and living areas. Their widely publicized capability to detect illegal drugs deters drug use and possession and is a valuable adjunct to a commander's other primary tools such as urinalysis and investigations.

Explosive Detection
MWD teams specially trained in explosive detection are exceptionally valuable in antiterrorism operations, detection of unexploded ordnance, and bomb threat assessment. Units will refrain from public demonstrations of their explosive detector dog capabilities.

Contingency Operations
In war fighting roles, MWD teams are tasked to bring enhanced patrol and detection abilities to perimeters, traffic control point (TCP), cordons, dismounted combat patrols, and point defense in bare-base operations; they are a rapidly deployable and effective sensor system. A well-trained and effectively placed MWD team can augment and enhance current contingency operations.

Perimeter defense

An MWD team can detect intruder(s) several hundred yards out from the team's position. Teams should be posted in such a manner as to allow the MWDs senses to be effectively utilized.

TCP operations

MWD team's explosive detection capabilities make them a necessity to all TCP operations. MWD teams participating in TCP operations will remain out of sight in a vehicle or building until needed. Choke points and search area should be placed in such a manner as to allow the MWD team to safely maneuver around the vehicles. Once vehicles are chosen, all personnel will exit the vehicle and all compartments will be opened for inspection by the team. Driver and any passengers should be removed or turned away from the immediate area while the team is searching and should not be allowed to observe the team while searching. The MWD team will have an over watch at all times.

Flash TCP operations

Flash TCP operations are the same as TCP operations, but MWD team is a member of a convoy/mounted patrol. MWD teams will remain out of sight within assigned vehicle until needed. Once vehicles have been chosen, a mounted patrol will entrap the vehicles and safely get them stopped and positioned in such a manner as to allow the MWD team to safely maneuver around the vehicles. Once vehicles are stopped, all personnel will exit the vehicle and all compartments will be opened for inspection by the team. Driver and any passengers should be removed and turned away from the immediate area while the team is searching and should not be allowed to observe the team while searching. The MWD team will have an over watch at all times.

Aerial TCP operations

Aerial TCP operations are generally the same as a flash TCP operations, but the MWD team is part of a squad traveling and conducting TCPs by helicopter. Once vehicles have been chosen, and stopped, get all personnel out of the vehicles and all compartments will be opened for inspection by the team. The MWD team will be the first on and last off of the helicopter. The MWD team will remain out of sight until needed. Driver and any passengers of the vehicle being searched should be removed from the immediate area while the team is conducting the sweep. The MWD team will have an over watch at all times.

Dismounted combat patrols
The MWD team and security member(s) will be located in the middle of the patrol when traveling within villages. One of the security member's purpose is to mitigate any threat to the MWD team during the combat patrol because the handler's attention must remain on the dog making him unable to scan for threats. MWDs can be positioned in front of the patrol if in an open field, walking from downwind if at all possible. This will maximize the MWD team's explosive detection capabilities.

Cordon and search (raids) operations
Initial entry team will clear all occupants from building as quickly as possible. Security member will escort the MWD team to the identified building. When the facility is cleared of personnel and hazards have been removed and/or identified, the handler will initiate a search of the facility.

Physical Security
Physical security augment in detection roles, replace inoperative sensor systems, patrol difficult terrain, and deter potential aggressors. Depending on what particular role MWD teams will serve determines the manner in which teams are posted. If the dog team's primary function is to provide direct security over the resource, its close proximity may limit the team's ability to follow up on any alerts the dog gives its handler. If the dog team is to provide security from a greater distance than other SF patrols, their immediate response to the resource will be slower; therefore the team should not be factored into any time-sensitive initial response requirements. Remember, to gain the maximum use of the dog team's capability, handlers must be able to work their animals in such manner so as to take full advantage of the dog's keen sensory abilities.

UNDERSTANDING MWDS

Advantages

MWDs have distinct advantages over a lone SF member. MWDs have superior senses of smell, hearing, and visual motion detection. The MWD is trained to react consistently to certain sensory stimuli—human, explosive, drug—in a way that immediately alerts the handler. The MWDs reaction to this stimulus is always rewarded by the handler, which reinforces the MWDs behavior and motivates the MWD to repeat the actions. People react to what they "think" a stimulus means. MWDs simply "react" to the stimulus and let the handler decide what it means.

Superiority of Senses

Though hard to quantify under almost any given set of circumstances, a trained MWD can smell, hear, and visually detect motion infinitely better than security forces personnel and, when trained to do so, reacts to certain stimuli in a way that alerts the handler to the presence of those stimuli. It is important to remember that MWDs are "biological" pieces of equipment having good and bad days, which is why training is crucial to their proficiency. Therefore, the continuous training of an MWD team must be kept at the forefront of any SF operation. Without proper training a dog team's capabilities will quickly diminish. Eliminating the "bad" days is important to the success of an installation's MWD program, which is why every level of command's participation is paramount.

Evaluation of Desired Tasks

The MWD can enhance operations throughout the entire spectrum of security forces roles and missions. The most important question a supervisor should ask is, "Where should we post the MWD to enhance the mission?" You have a dynamic force multiplier that tremendously enhances an SF member's ability. They have an asset that smells, hears, and detects better than anything else on flight. If this asset is left at the kennels, mission degradation is dramatic. The most important considerations are tasks required, the time of day to use the team, and the post environment. Consider these tasks:

Deterrence

If the desired task is to deter unauthorized intrusion, vandalism, attacks on personnel, etc., use the team on a post and at a time of day when those you wish to deter can see the MWD. MWD demonstrations are also beneficial in accomplishing this as well. People do not know if an MWD is a patrol dog, detector dog, or both. This is one benefit of public visibility of an MWD. Security forces benefit from the deterrence effect of every type of dog we train based on the presence of one MWD.

Detection

If the desired task is to detect unauthorized or suspect individuals, assign the team to a post at a time of day when visual, audible, and odor distractions are at a minimum. Examples include the flight line when operations are minimal; nuclear weapon storage areas; convoy operations and walking patrols in housing, shopping, or industrial areas after normal duty hours; WSAs; and other priority restricted areas.

Narcotic detector dogs (NDDs)/Explosive detector dogs (EDDs)

NDDs and EDDs are trained to detect specific substances under an extremely wide range of conditions, which makes post selection and time of day less critical.

THE MWD SECTION

The base CSF develops and is responsible for the MWD program supporting the installation.

MWD Logistics

Air Force Instruction 23-224(I), DoD Military Working Dog Program, sets policies and procedures governing logistical aspects of the USAF MWD program. It assigns responsibilities for budgeting, funding, accounting, procuring, distributing, redistributing, and reporting of MWDs and specifies procedures for submitting dog requirements and requisitions. When preparing and submitting MWD requisitions, consult with your installation logistics readiness squadron for assistance.

Obtaining Support Equipment

Use AF Form 601, Equipment Access Request, to order equipment to support the MWD program. A leash, choke chain, collar, and muzzle are shipped with the MWD to the gaining unit. Units must order other support items through supply channels or through use of the Government Purchase Card. SF units are authorized to pursue purchases of MWD support items outside the AF Form 601 channels, providing they comply with all USAF budgeting and requisition policies and guidelines. Consult with your unit resources advisor first.

MWD Section Organization

Most MWD sections are authorized a kennel master, a trainer (if five or more MWDs are assigned), and enough MWD handlers to meet the patrol standard of MWD teams. Refer to AF 31-202, Military Working Dog Program, for additional guidance.

Duties and Responsibilities

The basic organizational structure of an MWD section consists of a kennel master and a trainer.

Kennel master

The kennel master exercises management and supervision over the MWD program. The kennel master reports to the operations flight. The kennel master will:

1. Know unit mission.
2. Match dog to handler.
3. Assist in identifying MWD team posts and prepare operating instructions for team employment.
4. Ensure an adequate MWD training program is developed, implemented, and maintained.
5. Validate proficiency of MWD teams and prepare dog teams for certifications.
6. Ensure the health, safety, and well-being of MWDs are maintained by working closely with military veterinarian.

7. Ensure handlers understand basic principles of training and conditioning, physical and psychological characteristics, and the capabilities/limitations of their MWDs.
8. Obtain equipment and supplies needed for the unit's MWD program.
9. Advise the commander on effective MWD utilization.
10. Ensure unit and flight-level supervisory personnel are familiar with proper MWD team utilization and employment standards.
11. Perform duties as trainer if fewer than five MWDs are assigned.
12. Assume duties as primary custodian for narcotics and explosive training aid accounts.
13. Ensures handlers are properly trained on safety and security procedures associated with narcotics and explosive training aids.

Trainer

The trainer is directly responsible to the kennel master for managing and implementing an effective MWD training program. The trainer must be capable of performing all kennel master functions when necessary. The trainer should:

1. Schedule and conduct daily proficiency training following established optimum training schedule (OTS) requirements.
2. Schedule and conduct periodic intensive or remedial training for teams with special problems and training deficiencies.
3. Identify and correct deficiencies of handlers and MWDs in all phases of MWD operations.
4. Ensure MWD records are current and accurate.
5. Act as alternate custodian for the narcotic and explosive training aids.

Kennel support

Kennel support personnel need not be qualified handlers, although it is desirable. If not qualified, the kennel master must make sure support personnel are given local training in MWD care and feeding, kennel sanitation, disease prevention, symptom recognition, kennel area safety, and first aid/emergency care. Kennel support personnel who are not qualified handlers will not handle MWDs. Do not assign personnel who have been relieved from duty for "cause" as kennel support.

MWD handler

MWD handlers are SF personnel trained to use a specialized piece of equipment. Because of the time and effort required to keep a team proficient, employ the handler and MWD as a "team" and assign appropriate posts and duties. Avoid posting an MWD handler without his dog except for medical reasons. While unit manning shortfalls may require this as a last resort, keep it to a bare minimum, as it could rapidly create an adverse effect on MWD proficiency. **NOTE:** SF supervisors with dog teams in their charge must coordinate, with the MWD supervisory staff, MWD training and kennel well-being, and sanitation duties handlers are required to perform while on duty.

Accurately document all training and utilization of assigned MWD as directed by the kennel master and trainer.

CHAPTER 8

ADMINISTRATION/MEDICAL RECORDS, FORMS, AND REPORTS

ADMINISTRATIVE RECORDS, FORMS, AND REPORTS

The MWD staff maintains the following:

MWD Training Record Folder
A repository for MWD training records contains the following documentation:

DD Form 1834, MWD service record
Initiated when the MWD is first procured and entered into the DoD MWD inventory and kept current by kennel master throughout the MWD's service life. Annotate unit of assignment as well as assignment of new handlers on the reverse side of the form. Do not change information pertaining to MWD's national stock number without prior coordination with the MAJCOM and the 341 TRS.

LAFB Form 375, MWD status report
This form provides a record of training on the MWD after it graduates from a course.

AF Form 321, MWD training and utilization record

AF Form 321 provides complete history of patrol training, utilization, and performance. Handlers annotate each duty day and sign at the end of each month. The kennel master, as the reviewing official, will also sign it at the end of each month.

AF Form 323, MWD training and utilization record for drug/explosive detection

This form is a record of training, utilization, and performance of detector dogs. It serves as the basis for establishing probable cause. Annotate and sign the same as the AF Form 321.

OTS

Used to outline training requirements for each individual MWD, the OTS should concentrate on tasks that each particular MWD is trained to perform. Adjust OTS as a team's performance improves or deteriorates. If the team performance deteriorates, consider increasing training requirements within that area. If the team performs with little or no difficulty, consider decreasing training requirements in that area. Once a schedule is established, follow it! Place OTS in the MWD training records. Document training records to reflect any deviation from the OTS and the reasons why.

Controlled Substance Accountability Folder

The controlled substance accountability folder is used to provide a record of accountability for controlled substances. A separate folder is established for each substance and kept active until all controlled substances from that shipment are returned for final disposition. Once all substances from that shipment are returned, the folders are placed in an inactive file and retained for one year. The controlled substance accountability folder consists of the following documentation:

DEA Form 225, application for registration

The person assigned direct responsibility for control and safekeeping of narcotic training aids signs as the applicant. Refer to Title 21, Code of Federal Regulations (CFR), Part 1300, for specific details. DEA Form 225 is only needed for CONUS, Hawaii, Guam, and Puerto Rico SF units.

DEA Form 225a, application for registration renewal (type B)

Form 225a is required to maintain DEA registration. The form is mailed directly to the unit approximately 60 days prior to expiration of current registration. DEA Form 225a is only needed for CONUS, Hawaii, Guam, and Puerto Rico SF units.

DEA Form 223, controlled substances registration certificate

Valid for 1 year, unless withdrawn sooner by DEA, DEA Form 223 is only needed for CONUS, Hawaii, Guam, and Puerto Rico SF units.

DEA Form 222, controlled substance order form (type B)

This is an accountable form used to order drug training aids from the drug distribution center. Forward copies one and two to the drug distribution center; the unit maintains copy three. Upon receipt of the drug training aids, annotate the number of training aids and date received. DEA Form 222 is only needed for CONUS, Hawaii, Guam, and Puerto Rico SF units.

AF Form 1205, narcotics training aid accountability record

All SF units possessing narcotic training aids will record and account for these items using this form regardless of whether they are registered with DEA or not.

Drug Training Aid Issue/Turn-in Log

This log is used to document the issuing and return of MWD drug training aids. AFI 31-202, Military Working Dog Program outlines specific instructions on annotating the Drug Training Aid Issue/Turn-in Log. CSF authorizes, in writing, individual access to the drug storage container and those personnel who are authorized to possess drug training aids. This authorization letter must be posted within the area/room containing the storage container, but not on the storage container itself. Primary custodian must ensure personnel who are authorized to sign for and/or possess drug training aids are properly trained on their control and security. Document this training in the person's OJT records. For those who do not have OJT records (MSgt or above), document their training via official memorandum. Primary custodian will maintain a record of those personnel who have been trained and whether the individual reports to the kennel master or not. Only MWD personnel who have attended the MWD supervisor's course will be assigned duties as primary training aid custodian.

Probable Cause Folder
This folder is used to provide search granting authority an overview of the detector dog team's performance. Probable cause folders should reflect only the current team's performance. Maintain AF Forms 321 and 323 in the probable cause folder for 12 months. Upon removal, place AF Forms 321 and 323 in MWD training record folder. The probable cause folder should consist of the following documentation and provided to the search granting authority for review on a quarterly basis.

Search granting authority record review sheet
The search granting authority signs and dates a signature page certifying concurrence with the contents of the probable-cause folder and recertification of the team. Conduct this review/recertification quarterly within the calendar year.

Certification letter
This document discusses details of the certification demonstration to include search granting authority, or designee, witnessing of the demonstration. Conduct the certification annually as needed and when handler change occurs.

Quarterly summary report
This quarterly report details training and actual search utilization conducted by the team during the previous calendar quarter. The training summary consists of a breakdown of detector dog training by times, numbers of training aids planted and found, and by areas that training was conducted. The actual search summary consists of search times in each area searched including all non-training aid substances found by the dog.

Validation results
This is a summary of results from validation testing.

MWD team's resume of training and experience
This resume is a summary of training and experience for both the handler and MWD.

Medical
The servicing veterinarian maintains all MWD medical records. Only veterinarian staff personnel make annotations to medical records. Medical records are made available for deployments. Servicing veterinarians are responsible for initiating the following forms:

DD Form 1743, death certificate of military dog
DD Form 1743 is required for the death of all MWDs and includes a brief statement identifying the cause of death. It is used to close out accountability for a MWD through the base supply system.

DD Form 2209, veterinary health certificate
This certificate should accompany the MWD during interstate travel, depending on state requirements, or to foreign countries. Required for all commercial and military travel and may be required for military airlift. Certificate is valid for 10 days from date issued. The kennel master must ensure all health, customs, and agriculture requirement associated with MWD travel to foreign countries are satisfied prior to allowing MWDs to depart CONUS. Should problems arise in meeting a foreign country's animal entry requirements contact your AF MWD program manager immediately. For missions requiring a US Department of Agriculture health certificate, consult your supporting veterinarian for guidance. Your local travel management office is another good source for interstate and foreign travel requirements of animals.

Forms
The following are forms used within the MWD program:

AF Form 68, munitions authorization record
Used to document approval to procure explosive training aids, AF Form 68 is updated every 6 months or whenever changes occur. Keep the form current to procure explosive training aids. Refer to local munitions

account supply office (MASO) personnel for further information concerning completion of the form. Maintain most current AF Form 68 within the munitions accountability folder. Your local MASO provides guidance on maintaining this folder.

AF Form 321, MWD training and utilization record
This form is used to document the MWDs patrol training and utilization.

AF Form 323, MWD training and utilization record for drug/explosive detector dogs
This form is used to document the MWDs detection training and utilization.

AF Form 324, MWD program status report
The MWD Program Status Report is prepared by units and used by the USAF program manager and HQ AFSFC/DOD MWD program management office to effectively manage the USAF and DOD MWD program(s). Units provide original copy of report to USAF program manager quarterly, which in turn forwards reports to HQ AFSC/DOD MWD program management office. MWD status reports are due to the USAF MWD program management office NLT the 5th of the month following the quarter.

AF Form 601, equipment action request
AF Form 601 is used for major equipment acquisition to include MWDs. Contact unit resources personnel for assistance in completing this form. Consult AFI 23-224(I), DOD Military Working Dog Program and AFMAN 23-110/USAF Supply Manual, Vol 2, Part 2, Chapter 22.138 when submitting acquisition request for dogs.

AF Form 1996, adjusted stock level
Completed on an annual basis, this form is used to establish yearly levels and re-supply increments for explosive training aids. Due to sensitivity and transportation factors, accomplish a 5-year forecast for explosives. Consult with your local MASO and AF Catalog 21-209, Volume 1, Ground Munitions, Chapter 1.

AF Form 2005, issue turn-in request

AF Form 2205 is used for small, expendable items that do not require approval above base level. Maintain all pending action copies of AF Form 2005 in munitions accountability folder along with other related accountability documents. Your local MASO provides guidance on maintaining this folder.

DD Form 1348-6, DoD single-line item requisition system document

This document is used for local procurement of munitions/equipment items when no national stock number (NSN) is available. Consult your local MASO prior to using this form for munitions requisition.

DD Form 2342, animal facility sanitation checklist

Completed on a quarterly basis, this form includes standard of sanitation maintained, the adequacy of insect and rodent control, and the general health of MWDs judged by their appearance and state of grooming. The CSF reviews and signs completed forms that will be maintained at the MWD section.

CHAPTER 9

FACILITIES AND EQUIPMENT

KENNEL FACILITIES

Some existing facilities may not meet current construction guidelines. If they conform to health and safety requirements, they do not require modification. See the Kennel Design Guide for guidance regarding any new kennel construction located on the Headquarters Air Force Security Forces Center website (afsfmil.lackland.af.mil) under the DoD MWD Program Management Branch tab.

Kennel Maintenance

MWD supervisory staff will inspect kennel facilities and dog runs continuously to ensure safety and security of MWDs and personnel. Inspect all latches, hinges, and fences for signs of rusting or breakage. Free all surfaces of sharp objects that could cause injury.

Sanitation Measures

Sanitation is one of the chief measures of disease prevention and control and can't be overemphasized. MWD supervisory staff in conjunction with supporting veterinarian establishes and enforces stringent kennel sanitation standards in and around the kennel area. An effective, continuous sanitation program is the result of cooperation between handlers, supervisors, kennel support personnel, and the attending veterinarian. The kennel master, with consent of supporting veterinarian, approves all cleaning products and solutions used to sanitize kennel runs and dog feeding preparation area.

Food Preparation and Storage

Keep all kitchen surfaces and food preparation utensils clean at all times. Store dog food in rodent proof containers with excess food awaiting use stored off the floor. Ensure new food bags/containers are inspected before feeding. Do not use feed from containers if the manufacturer's product packing seal is broken or punctured. Dispose of uneaten food immediately after the feeding period. Empty all trash containers as needed, at least daily, to preclude attracting pests into the facility.

OBSTACLE COURSE

Construct the obstacle course in accordance with guidance provided by HQ AFSFC. Variations in construction material are authorized with attending veterinarian concurrence. Cover surfaces with nonskid material and cover all sharp edges. The kennel master ensures the obstacle course is maintained in a safe condition consistent with guidance from the servicing veterinarian. All MWD sections will have a serviceable obstacle course.

AUTHORIZED EQUIPMENT

Kennel masters must ensure all equipment is available and serviceable. Kennel masters may establish local purchase programs through base supply to acquire additional equipment.

Choke Chain

The choke chain is the basic collar used for all MWDs.

Leather Collar

Use the leather collar when securing an MWD to a stationary object (stakeout). Tighten the collar to the point that the handler can slip two fingers snugly between the collar and the MWD's neck.

Kennel Chain

Use the 6-foot kennel chain with the leather collar when securing the MWD to a stationary object. Attach the kennel chain to the D-ring of the collar with the snap facing away from the buckle. Never tie/loop the kennel chain around the MWD's neck.

Muzzle

Use the leather muzzle, safety muzzle, or suitable plastic replacement to prevent the MWD from injuring the handler, other MWDs, and people. Use muzzles during veterinary visits or first aid treatment, and when numerous MWDs are assembled, in transit, or in crowded confined areas. When properly fitted, the muzzle will not restrict breathing. Check the fit by grasping the basket of the muzzle and lifting straight up until the MWD's front feet are off the ground. If the muzzle comes off, adjust accordingly. MWDs supporting U.S. Secret Service, Department of State, and/or Department of Defense explosive detector dog missions will be muzzled when traveling (on foot) to and from search locations or any other public area not directly associated with search activity.

Leashes

The 60-inch leather leash is the standard leash for MWD operations. Use the 360-inch nylon or web leash for intermediate obedience, attack training, and tracking operations. Kennel masters may approve other leashes, such as heavy duty retractable models, to meet operational requirements. Because the leash is the most used piece of MWD equipment, handlers will inspect the serviceability of their leashes daily. **DO NOT WORK MWDs WITH NONSERVICEABLE OR DEFECTIVE LEASHES.**

Equipment Holder

Use the equipment holder to secure items to the handler's belt.

Combs and Brushes

Use assorted combs and brushes to maintain high MWD grooming standards.

Feed Pan

Use the 3-quart stainless steel feed pan for MWD feeding.

Water Bucket

The water bucket is steel or galvanized metal and holds at least 3.5 gallons. Use feed pans for small breed MWDs. If this type of bucket is used, ensure drain holes are punched into the side of the bucket to prevent overfill. Punch drain holes approximately at the one-third full point of the bucket.

Immersion Heater
The immersion heater is automatic and has a thermostat to keep 12 quarts of water at 50°F. To work effectively, submerge in at least 2 inches of water. Inspect the power cord before each use. Don't use if the cord is unserviceable.

Leather/Nylon Harness
The MWD wears the harness while scouting or tracking. It enables the handler to control the dog's ranging distance but still allows the dog to breath normally. To fit the harness, place the leash in the left hand and the harness in the right hand, thread the loop end of the leash through the center of the harness. With the harness resting on the left forearm, change the leash to the right hand. Slide the harness over the MWD's head and shoulders and buckle the stomach strap behind the MWD's front legs. Then grasp the center of the back strap with the left hand, unsnap the leash from the choke chain and snap to the D-ring of the harness with the snap facing downward.

Arm Protector
The agitator uses the arm protector during aggression training. Use a leather gauntlet under the arm protector when training those dogs that bite hard.

Attack Suit
Consider using the full body attack suit acquired at the kennel master's discretion for advanced training. One major advantage of the suit is the ability to train dogs to bite and hold a suspect who does not present the attack sleeve. Some dogs become reliant on the sleeve and will not bite an actual suspect unless a target is presented. Using the suit properly will aid in correcting this behavior.

MAINTENANCE OF EQUIPMENT

Safety and extension of serviceability are primary objectives of all equipment maintenance plans.

Leather Items

Apply saddle soap or Neatsfoot Oil to preserve the strength of the leather and prevent drying or cracking. Ensure surfaces are clean and dry prior to application. Never apply Neatsfoot Oil to inside surfaces of the leather muzzle.

Metal Parts

Remove rust by rubbing with fine steel wool or sandpaper. Use a light coat of oil if necessary, replace badly rusted items, and inspect snaps routinely to ensure they are working properly.

Care of Fabrics

Wash web leashes with a mild soap and dry slowly to prevent shrinkage. Do not wash the arm protector with soap and water. When this item becomes dirty, clean by rubbing briskly with a coarse brush. To ensure safety, frequently check the arm protector. Make minor repairs with a needle and heavy thread.

Storage of Equipment

Keep all equipment dry when not in use. When in storage, inspect and treat as needed to ensure that it is clean, soft, and in good condition.

VEHICLE AUTHORIZATION FOR KENNEL SUPPORT

Kennel support requires a suitable vehicle for transporting explosives as well as MWD teams.

SHIPPING CRATES

Shipping crates are authorized for each MWD in a unit type code (UTC). Smaller airline approved plastic shipping crates are also authorized to support TDY's requiring transport via commercial aircraft or ground transportation. Plastic commercial shipping crates are not to be used for long-term kennels. Refer to applicable UTC LOG DET.

CHAPTER 10

SAFETY AND TRANSPORTATION PROCEDURES

KENNEL SAFETY

Following sound safety procedures in kennel and training areas is very important. Personnel must follow safety practices at all times. Maintain positive control or a dog may get loose and injure a person or itself. Safety practices begin as soon as a person enters the kennel area. Personnel must ensure they secure all gates after use, avoid sudden movement when passing MWDs, and not speak or move in any threatening way. Personnel must not run or horseplay in or near MWDs. This activity agitates MWDs and could result in a dog mistaking it for hostility and provoke attack and cause injury to the dog.

One-Way System
Set up one-way traffic patterns in kennel areas to keep dogs from meeting head-on. Ensure the one-way system is clearly marked.

Loose Dog Procedures
If a dog gets loose, the first person observing the MWD calls out "LOOSE DOG!" Everyone except the handler should cease all movement until the dog is secured. Once the MWD is under control, the handler must sound off with "DOG SECURED!" SF units must develop local procedures to protect the public should an MWD escape the kennel area.

Verbal Warnings

Handlers with MWDs will give verbal warnings upon entering or leaving the kennel area or when vision is obstructed by calling out "DOG COMING THROUGH, AROUND, BY," whichever is appropriate.

Dog Fight Procedures

If a dog fight occurs, never attempt to stop it alone and never pull MWDs apart. Pulling may cause greater damage. If on-leash, keep the leash taut and work your hands toward the snap of the leash. Hold the leash firmly with one hand, grasp the MWD's throat with the other hand, and squeeze with the thumb and forefinger to cut off the air supply. When the dog gasps for air, move it away. If off leash, grasp the choke chain, leather collar, or nape of the neck with one hand. With the other, squeeze the dog's throat using the thumb and finger to cut off the air supply.

TRAINING AREA

The following safety precautions are required in the training areas:

1. Keep a safety leash on the right wrist while moving to and from training areas.
2. Keep a minimum safety distance of 15 feet between MWD teams in the areas. When approaching another MWD team, keep dog in the heel position using a short leash.
3. Never use a leash to secure an MWD to any object. Never leave an MWD staked out unobserved, and never secure an MWD to a vehicle.

SAFETY IN THE VETERINARY FACILITIES

When a MWD is taken to the clinic, it is around unfamiliar surroundings and people and may behave unexpectedly. The handler must control the dog while at the clinic. Get clearance from veterinary staff prior to entering the clinic.

1. Before entering the veterinary clinic, the handler will muzzle the MWD, unless instructed otherwise by the veterinarian staff.

2. The handler must give a verbal warning "DOG COMING THROUGH" before entering. In the treatment facility, the handler controls the MWD with a short leash.

OPERATIONAL SAFETY

Safety considerations are of paramount importance. Apply safety practices at all times for the protection of the handler, other handlers, MWDs, and the general public. When dog teams interact with the public, handlers must be in the mindset that it is the handler's responsibility to be safe around the public, not the public's responsibility to be safe around the dog team.

Static Posts
While working a static post, handlers must remain aware of their surroundings. Do not allow anyone to pet their MWD.

Mobile Patrol
While riding in a vehicle, avoid sharp turns and sudden stops whenever possible because they could result in injury to the MWD. Train MWDs not to attack personnel riding with the team. Do not allow the dog to ride with its head outside the vehicle window. Do not leave MWDs unattended in a vehicle except in an emergency situation that necessitates the handler responding without the MWD. However, if you must leave an MWD in a vehicle, ensure the vehicle is secure so the MWD cannot escape. Also, ensure proper ventilation so the animal does not overheat. The handler must have full view of the vehicle at all times. This ensures the handler can assist the MWD if it becomes distressed.

DoD and Civilian Law Enforcement Support Agency Operations
The same principles of safety that apply when using MWDs on the installation also apply when deployed in support of outside agencies. Handlers and trainers must remain aware of the potential danger of MWDs that are trained to attack. Whenever necessary, advise personnel on the safety procedures.

VEHICLE TRANSPORTATION

Use the hindquarter or abdominal lift when loading MWDs on a vehicle. To place an MWD in a vehicle for patrol purposes, begin with the MWD in the HEEL position. Open the door and command "HUP" and then "SIT."

AIRCRAFT TRANSPORTATION

Use commercial and military aircraft when shipping MWDs interstate or to overseas commands. Do not route MWDs through countries with quarantines. Consult the TMO office for details.

Commercial Air Transportation

When MWDs are shipped unaccompanied, attach detailed instructions for feeding and watering to the crate. Mark the top of the crate with the name and tattoo of the dog as well as special instructions noting the kennel should only be opened by a qualified handler or not at all. Specific instructions on contacting a point of contact for the dog will also be noted in case of emergency. Verbiage should be large enough to read from a safe distance. The top and sides of the crates should be labeled "DANGER—MILITARY WORKING DOG."

1. Stay with the MWD until loaded. If there is a delay, remove the dog from the crate for exercise and water.
2. Always place the crate in a cool spot when waiting for loading. Unload the MWD as soon as possible and make sure it has water.
3. Never place the crate on top of other baggage.
4. Do not lock shipping crates! They must be able to be opened in an emergency. Do make sure, however, they cannot be opened inadvertently.
5. If shipped accompanied, ship the MWD as excess baggage. Check with the local carrier for an exemption to the excess baggage fee. Refer to AFI 31-202, paragraph 2.7.5. for further guidance.

MILITARY AIR TRANSPORTATION

MWD handlers are required to escort MWD movements on military aircraft. Escorting handlers can make recommendations to aircraft crew personnel on the best way to load the dog, but ultimately the aircrew will have the final decision. Ensure, however, that load plan does not block ventilation to the crate; furthermore, cargo and other equipment should not be placed on top of the crate. The handler must have ready access to the crate door in the event of dog distress or injury.

EXPLOSIVE SAFETY

Refer to AFMAN 91-201, Explosive Safety Standard and AFI 31-202, *Military Working Dog Program.*

DRUG SAFETY

If an animal ingests a training aid, contact the veterinarian immediately to determine whether the animal was poisoned and what actions to take to ensure the dog's safety.

CHAPTER 11

OPERATIONAL EMPLOYMENT

SECURITY OPERATIONS

Mission

The MWDs primary mission is to deter, detect, and detain intruders in areas surrounding Air Force resources. Use MWD teams on almost any security post. Include the kennel master in planning use of MWDs in security operations.

The greatest advantage of an MWD team is their detection capabilities and their ability to cover a large area, particularly during periods of limited visibility. The presence of an MWD team may also discourage attempts by intruders to gain access to resources. MWDs detection capabilities are degraded in areas with a large number of people and constant activity. When used on a post where there is little room to maneuver to take advantage of wind direction, the MWD must depend mostly on its sense of sight and sound. When working on a post where the handler must concentrate on tasks other than working the MWD such as entry control, the MWD's abilities are largely wasted.

Post Selections

Ensure MWD posts are free of distractions such as excessive noise and numerous personnel. If deterrence is the objective for assigning a MWD team to a particular post, you will sacrifice detection ability for the higher degree of visibility. Wind direction, the location of priority resources, size of the area, condition and type of terrain, and likely avenues of approach dictate the best location for an MWD post. Under ideal conditions, the average MWD can detect and respond to intruders at 250 yards or more. Make efforts to keep MWDs working downwind, because this is where their sense of smell and hearing is best used. Keep post selection and limits flexible enough to meet varied conditions. If you must use an MWD team in a lighted area, allow it to patrol on a more varied route, remain in the shadows, or stand stationary in a concealed downwind position. Using MWD teams in lighted areas reduces the team's ability to remain undetected. This permits intruders to observe their movements and increases the possibility of successful penetration. Also, lights may cause the MWD to rely more on sight than its other senses. Because there are no steadfast rules on the number and location of MWD posts, make post selections with common sense in relation to environmental factors.

Close boundary (CB) post

An MWD team on CB foot patrol provides security that far exceeds the capabilities of the lone security force's member. This is especially important when considering a sentry's effectiveness is limited by darkness due to poor visibility. The MWD team's objective is to detect and apprehend intruders before they can damage or destroy the protected resources. Consider these other factors:

- Posting the MWD team inside a fenced area allows the team to patrol close to the resource, physically checking it periodically. However, if the MWD detects an intruder attempting entry, the fence will prevent the team from following in the response, possibly preventing or delaying apprehension. If an additional MWD is not available, take the MWD team with the response outside the area and allow it to follow up the response. Personnel posted inside the area should increase their vigilance as well as observe the team's post until the situation returns to normal.

- Posting MWD teams outside a protected area creates the advantage of using the wind, cover, and concealment, and the opportunity to follow responses to the source. The team can effectively cover approaches to the area and detect odors from within the area by using the wind to their advantage.

- Give consideration to using a flexible posting system. This permits using the MWD in or out of the area to meet varying conditions and increases psychological protection by preventing a routine patrol pattern.

Supplementing Intrusion Detection Equipment (IDE)

When planning MWD posts, consider using electrical or mechanical detection devices. Because these devices usually activate only when detecting intruders within the surveillance area, use the MWD more effectively as a backup for these systems.

Team position

Position the team inside the protected area so it can patrol the whole area around the resource without setting off an alarm. The MWD handler should maintain close contact with the alarm monitor. The MWD team can then approach from the downwind side and make contact with the intruder.

Mobile Security Patrols

Vary duties of a mobile patrol to include building checks, area surveillance, and identification and apprehension of personnel. MWD teams on internal or external security response teams (SRTs) are effective force multipliers, cover large areas, and present both a physical and psychological deterrent.

Response Forces (RFs)

Enhance the effectiveness of an RF by using an MWD team during open-area searches, scouting, tracking, building searches, and apprehensions. The MWD is an integral part of the team, and you should not separate it from other members. Familiarize all RF members with the MWD's capabilities and procedures to follow. The team must discuss the situation and decide upon the approach route and what to do upon arrival. There is no set rule for deploying an RF area patrol when an MWD is used; therefore, circumstances will dictate each response. If the team leader decides not to use the MWD team in the deployment, he should have the handler remain with the vehicle and operate the radio.

Explosive detector dogs

After the situation is neutralized and declared safe, use explosive detector dogs to sweep areas for unexploded ordnance, explosive devices, weapons, and ammunition.

AIR PROVOST OPERATIONS

Psychological Impact

MWD teams in law enforcement activities offer a tremendous psychological deterrent to potential violators and should work in all areas of the base. Psychological advantage is complemented by conducting periodic public demonstrations. Keep these demonstrations as realistic as possible and include obedience, attack under gunfire, and drug detection techniques. Using local news media and conducting special demonstrations are excellent ways to enhance community relations (both on and off base), local Drug Abuse Resistance Education (DARE) programs, and deterrence of unlawful acts on Air Force installations. Supervisors should limit the number of demonstrations—MWDs are working dogs and not show dogs. Discourage public demonstrations by explosive detector dogs as this may tend to generate prank or hoax bomb threat calls.

Use

MWD teams can perform in all SF functions. Include the kennel master in planning the use of MWDs in SF operations. It is the flight leader's/sergeant's responsibility to ensure MWD teams are posted in areas that will capitalize on their capabilities. Do not post handlers without their MWDs except in extremely rare situations (i.e., MWD is ill, back-up force response when time is of the essence). MWD teams will discourage unruly and/or unlawful conduct and increase the probability of apprehension. A properly trained MWD can pursue, attack, and hold a suspect providing an alternative to the use of deadly force.

Resources protection

Consider using MWD teams in AA&E storage areas during hours of darkness.

Military housing and billets

MWDs are especially effective in and around military housing and billet areas. Their mere presence can deter thefts, burglaries, drug use, and vandalism. Use teams both day and night in these areas.

Protection of funds

Using MWD teams to escort and safeguard funds may deter a robbery. An MWD does not fear an armed or unarmed person and, if fired upon, will pursue and attack—an important characteristic to emphasize during demonstrations and in news releases.

Confrontation management

Use MWD teams judiciously in confrontation situations because their presence could escalate the situation. Do not deploy MWD teams on front lines in riot control situations; keep them out of sight, and use as necessary.

Narcotics/explosives detection

A trained detector dog can detect drugs or explosives regardless of efforts to mask the scent. Publicity on the presence and effective use of drug detector dogs may help reduce criminal activities involving drugs. Do not make public information on limitations and effectiveness of a detector dog.

Patrols

Walking patrols

MWD teams should perform various duties such as checking buildings, parking lots, military housing, and billet areas. When used in this capacity, consider several factors:

- Use MWD teams as much as possible during both day- and nighttime hours in areas where one can easily see them. They should be tolerant of people, and the presence of crowds should not significantly reduce their usefulness. MWD teams should walk among people, stand guard mount, and show no outward sign of aggression. Extremely hot weather may restrict the MWD's abilities. In this case, divide the MWD team's time between mobile patrol and walking patrol.

- When working among the public, handlers should keep in mind the MWD is a valuable tool and not a pet. Maintain a safe distance and do not let anyone pet the MWD.

- The MWD's ability to detect individuals is more effective during darkness or limited visibility when there are fewer distractions. Therefore, give nighttime use in areas with few people, but high value resources, the highest priority.

- An MWD team can check or search a larger number of buildings and parking lots more efficiently than a single person.

- Periodic use of MWD teams around on-base schools (especially when school is starting and dismissing) may deter potential vandals, child molesters, exhibitionists, and illegal drug activities.

- Use MWD teams to provide security for such resources as aircraft, munitions storage areas, communications facilities, equipment, or command posts. When assigning walking patrols, one restriction to keep in mind is the lack of mobility. Like any security forces foot patrol, this could prevent a rapid response to a distant incident where time is essential.

Mobile patrols

Mobile MWD teams increase their potential area of coverage but decrease MWD effectiveness. Teams are usually unaccompanied; but because MWDs can work in the proximity of people, other security forces personnel may accompany them. Assign the mobile MWD team a sedan or other passenger-type vehicle—air conditioned in hot climates. If using a pickup truck, placing portable kennels in the beds of pickups for transporting MWDs while on patrol is prohibited.

- When mobile patrolling, allow the MWD to ride off leash. (If the leash is kept on, drape it over the dog's back to prevent it from getting caught on anything.) Use a vehicle kennel insert or a specially designed platform when an MWD is on mobile patrol. Cover the surface with rubber matting or some other nonskid surface. The MWD should remain in the "SIT" position as much as possible. Do not allow the MWD to place its head out of the window while the vehicle is moving.

- MWD teams should not remain mobile during the entire tour of duty. MWD patrols are more effective when the team uses the ride-a-while, walk-a-while method. The team is able to cover a larger area, and the exercise keeps the dog responsive.

Building checks and searches

MWD teams are especially effective in checking and searching buildings such as commissaries, base exchanges, finance offices, banks, and warehouses. With the MWD on leash, approach the building from the downwind side to take advantage of the MWD's olfactory senses. The responding patrolmen secure the surrounding area to avoid contamination by a fresh scent. This could serve to confuse the dog in the event tracking is required.

If a facility is found unsecure, the team should request backup. Upon arrival of the backup patrol, the MWD team should approach from the downwind side and first check the exterior before entering the building. The on-duty supervisor, after conferring with the MWD handler, determines whether the MWD should search on or off leash. Generally, the MWD is most effective when worked off leash because the dog's movements are not restricted, and it can search a larger area in a shorter period of time. Before a handler releases an MWD inside the building, announce in a loud clear voice the intention to release the MWD, and anyone inside the building should exit within a set period of time (1–5 minutes). Before releasing the MWD, the handler should consider the following factors: danger to the handler, type and size of building, time of day or night, indication of forced entry, and the possibility of innocent persons. The handler must check and clear the immediate area before proceeding. As the handler follows the MWD, they should use the same precaution for each room or area. One suggestion includes turning on lights as the handler progresses. However, keep in mind turning on lights will silhouette the handler when entering/exiting the room. Also, consider the consequences of turning on lights if there is an explosive device with a light sensitive switch in one of the rooms. If the MWD responds, it is recalled, placed on leash, and the intruder is challenged and apprehended. For an on-leash search, the handler enters and loudly announces that an MWD is being used to search. Another security forces member should always accompany the team. The assisting SF follows at a distance to avoid interfering with the search. If it is determined the suspect has exited the building, the MWD team should attempt to track the suspect from the scene. Tracking may result in additional evidence or information for a subsequent investigation.

Vehicle parking lots
Use MWD teams to detect and apprehend thieves and vandals in parking lots. The mere presence of the team may deter potential acts of theft and vandalism. The MWD team should approach from the downwind side.

Military housing and billet areas
Proper use of MWD teams in military housing areas will deter and decrease unlawful acts. During foot patrols, MWD team contact with area residents helps in the reinforcement of community relations. Outline clear procedures governing release of MWDs in military housing or billet areas in local operating instructions.

Alarm responses
Use the MWD team to search and clear the exterior and interior of alarmed buildings and surrounding areas. They may also assist in apprehensions. Limit the number of personnel allowed into the area to preclude contaminating the area with unnecessary scents.

Funds escort
When escorting funds custodians on foot, position MWD team slightly to the rear of custodian to observe any potential hostile acts. MWD handlers should brief fund custodians on actions to take during an attempted robbery. If local procedures allow the carrier to ride in the security forces vehicle, position the MWD where the handler will have positive control.

Moving traffic violations
When a traffic stop is made, the MWD should accompany the handler on leash. The presence of the MWD will convince most offenders to remain cooperative.

Identification and apprehensions
When conducting identification checks or effecting apprehension, the handler must inform the person(s) that any display of hostility could

result in the MWD biting without command. If an apprehension is made, conduct a search with the dog in the guard position. If available, use a backup to transport persons taken into custody. When circumstances require an MWD team to transport personnel taken into custody, and the vehicle is not equipped with a vehicle insert or platform, position the MWD between the offender and the handler.

Riot and Crowd Control

Normally, do not use MWDs for direct confrontation with demonstrators. In fact, the presence of MWDs could aggravate a situation. During the peaceful stages of a confrontation, hold MWD teams in reserve and out of sight of the crowd. If the situation deteriorates, move MWD teams up to within sight of the crowd, but still well away from the front lines. Only when actual physical confrontation erupts, give consideration to employing MWD teams on the front lines. Once committed, use MWD teams as a backup force, integrated into the front line of forces, or use to assist apprehension teams.

Employment

When engaged in direct confrontation, keep MWDs on leash and allow biting only under specific circumstances authorized by the on-scene commander. Position other riot control force personnel approximately 15 feet from MWD handlers. Do not release MWDs into the crowd.

In an open area, chemical riot control agents will not normally adversely affect the MWD's capability to act as a psychological or physical deterrent. However, handlers should watch their dog closely under such conditions. If an MWD shows any signs of distress, have it examined by a veterinarian as quickly as possible.

Support duties

In large areas such as open fields, position MWD teams on the outer perimeter to contain the crowd while other forces make apprehensions. Post MWDs around holding areas and processing centers to prevent the escape or liberation of prisoners. Use MWD teams to assist teams in apprehending and removing specific individuals within a group of demonstrators. In this role, use the MWD team to protect members of the apprehension team not to effect the apprehension. Exercise extreme caution in these situations. The MWD could become extremely excited and agitated and could mistakenly bite a member of the apprehension team. The handler must maintain positive control over the dog.

Civil Disasters
Provisions exist to provide MWD teams to a civilian community to assist in humanitarian or domestic emergency roles. For example, MWD teams may help locate lost children or search an area or building that has received a bomb threat. Exercise extreme caution in these situations to ensure the Posse Comitatus Act is not violated. Coordinate all requests for assistance with the base staff judge advocate. Refer to AFI 10-801, Assistance to Civilian Law Enforcement Agencies, for additional guidance.

Protecting Distinguished Visitors
Employ MWDs around quarters and conference locations or for searching and clearing buildings. Used as a foot patrol, the MWD can use all its detection senses.

Fixed Post (Stake-Out)
The primary function of an MWD team on a fixed post is surveillance over an area or building. If used outdoors, locate the team downwind where the dog can detect a person by scent. If this is not possible, locate the team where the MWD may detect by sound or sight. When used indoors, the MWD must rely primarily on its sense of hearing. Other security forces personnel may accompany MWD teams on fixed posts.

Installation Entry Control
Use of MWD teams as entry controllers for extended periods of time seriously degrades their operational effectiveness. If circumstances warrant the posting of the MWD team on a base entry control point, the duration of posting should be kept to a minimum consistent with flight manning. Posting of MWD handlers without their assigned MWD is misuse of assigned resources. While performing as entry controllers, the MWD's primary function is psychological deterrence and handler protection. Permit the MWD to sit or lie down, but do not confine where it can't respond when needed.

Confinement Facilities

The CSF must authorize the use of MWD teams to augment inmate control procedures. MWD teams may be used to search for escaped inmates and conduct facility contraband searches but will not be used to guard inmates. MWDs will NOT be used in the interrogation or interview of prisoners, enemy prisoners of war (EPWs), or detainees.

CHAPTER 12

CONTINGENCY OPERATIONS

MWDs ROLE IN CONTINGENCY OPERATIONS

An MWD's excellent sensory skills coupled with its psychological deterrence make it a vital part of base defense and force protection missions. When establishing ground defense operations, MWD teams should be used to enhance the detection capabilities of the ground defense force and provide a psychological deterrent to hostile intrusion. Properly positioned, MWD teams are capable of providing an initial warning to the presence of hostile intruders. Past experience has shown that MWD teams often provide warning of attacks early enough to allow response forces time to deploy and prevent enemy forces from reaching their objectives. MWD teams can also be used to clear protected areas of hostile personnel, explosives, and weapons after an attack as well as prevent the introduction of explosives to an installation.

BACKGROUND

History has shown the enemy strategy in guerilla or terrorist attacks against Air Force installations has been based on surprise and concentration of forces against weak points. The aggressors rely on advanced planning, preparation, concealment during approach, sudden attack, and swift withdrawal. Therefore, effective early detection and warning systems are critical. Use of MWDs has proven their effectiveness in helping fulfill this role and complement technology gaps. MWDs used on tactical perimeter posts can provide warning of an impending attack early enough to allow for the deployment of the response force, thus preventing the enemy from reaching their objective.

MWD ORGANIZATION

QFEBP

QFEBP consists of an NCOIC (must be a graduate of the MWD Trainer/Kennel Masters Course No. L8AZR3P0710K1A) deployed in support of 9 to 15 MWDs when there is no in-place kennel support at deployed locations.

Responsibilities

- Coordinate in the planning of defense and posting of MWD teams.
- Advise leadership and defense force personnel on MWD capabilities, limitations, and effective employment of MWD teams.
- Coordinate veterinarian support at deployed location. Train handlers on any unique health care concerns at deployment location.
- Coordinate with defense force commander (DFC) in planning kennel location.
- Establish kennels and support areas.
- Develop and implement SOPs for the MWD section.
- Conduct orientation training to evaluate MWD teams.
- Once the area is secure, conduct validation training immediately.
- Develop and implement a training program to maintain MWD teams' proficiency.

QFECP

QFECP consists of one MWD trainer who is a graduate of the MWD Trainer/Kennel Masters Course No. L8AZR3P0710K1A11.

Responsibilities

- The trainer is directly responsible to the kennel master for managing and implementing an effective MWD training program. The trainer must be capable of performing all kennel master functions when necessary.

- Schedule daily proficiency training following established OTS.
- Schedule and conduct periodic intensive or remedial training for teams with special problems.
- Identify and correct deficiencies of handlers and MWDs in all phases of MWD operations.
- Ensure MWD records are current and accurate.
- Act as alternate custodian for the narcotic and explosive training aids.

QFEBR
QFEBR consists of one MWD and one handler. All MWDs will be qualified patrol explosive detector dogs.

- The QFEBR can be tasked to perform duties as mounted/dismounted patrols, fixed positions, or perform explosive detection missions.

QFEDD
Consists of one MWD and one handler. MWD team is qualified only for explosive detection.

- The QFEDD can be tasked to perform duties based solely on its explosive detection abilities.

QFEND
QFEND consists of one MWD and one handler. All MWDs will be qualified patrol narcotic detector dogs.

- The QFEND can be tasked to perform duties as mounted/dismounted patrols, fixed positions, or perform narcotic detection missions.

QFEPD
QFEPD consists of one MWD and one handler. All MWDs will only be qualified patrol dogs.

- The QFEPD can be tasked to perform duties as mounted/dismounted patrols or fixed positions.

Chain of Command

The alignment of the MWD teams within the chain of command should be based upon how the DFC intends to utilize the MWD teams within the defense. Example: If MWD teams will be tasked with a variety of missions such as mounted/dismounted patrolling, detection searches, and other related duties, the centralized alignment would be beneficial to allow the kennel master to select the most suitable MWD team for each mission. At a deployed location where MWD teams will be doing primarily the same duties every day, such as explosive detection at ECPs, the decentralized alignment may be more effective.

Centralized alignment

The QFEBP reports directly to the S3/Operations. The assigned MWD teams reports directly to the kennel master. All requests for MWD support are made through the S3 who in turn coordinates with the kennel master. Once the MWD team is tasked to a sector, that sector maintains OPCON (operational control) over the MWD team while the team is assigned.

Decentralized alignment

Under this alignment, MWD teams are assigned directly to a sector command post and directly support sector defense operations.

PRE-DEPLOYMENT

Planning must be started upon initial notification of deployment. This planning must include acquiring needed equipment, veterinarian support, how teams will be transported to the deployed location, kenneling at enroute stops and final destination, and how the teams will be employed in the deployed area of responsibility.

Administrative

All MWD handlers will deploy with:

• Copies of the most recent four months training records.
• Most recent validation letter.
• Copy of the MWD's health records.

Equipment

A number of considerations must be taken into account when planning the equipment requirements for the deployment. The length of the deployment, terrain and weather at the deployed location, as well as the amount of time it will take to establish resupply channels. A minimum of 30-days' supply of food and medication must be taken on all deployments. If the MWD team deploys with a LOGDET (QFE4R), some items may need to be added depending on the terrain and weather at the deployment location. An example of this would be box fans for a hot climate or cloth booties for the dog's paws in rugged terrain.

Transportation

The mode of transportation will determine the type of kennel that will be used for shipment. Unless specifically tasked otherwise, when traveling by military aircraft, teams will deploy with a suitable approved portable kennel. When traveling by commercial aircraft, the plastic vari-kennel type container will be used. When deploying to a location that does not have a finished kennel facility, arrangements must be made to have suitable kennel crates shipped to the deployed location to allow for the establishment of a field kennel. Plastic transport kennels are unsuitable for temporary kenneling and will not be employed in this manner. The home station kennel master is responsible to check for any quarantine requirements at stops enroute to the deployed location.

Veterinarian Support

Veterinarian support will be required prior to deployment, during deployment, and upon redeployment. Upon initial notification of deployment, the kennel master must coordinate with the home station veterinarian to get health certificates issued; copies of the MWD's health records to deploy with the MWD teams and identify any medical threats to the MWD at the deployed location as well have an MWD first aid kit unique to the deployed location prepared. The home station kennel master is responsible to ensure that the deploying handler is competent in emergency first aid and MWD life-saving skills. Prior to deployment, the home station veterinarian will be consulted as to the need of any insecticides that should be acquired to treat the area for insects such as mosquitoes and ticks. The handler must know how and when to use items deployed in the first aid kit. It is the deployed Kennel Master/ senior handler's responsibility to coordinate veterinarian support at the deployed location.

Kenneling
The type of base the teams will be deployed to will determine the type of kennel used: main base, stand-by base, or bare base.

Main base
A main base will have adequate facilities in place for maintaining MWD operations.

Stand-by base
A stand-by base or enroute stop may or may not have adequate facilities for maintaining MWD operations. In these cases a temporary field kennel may have to be established.

Bare base
A bare base will not have any facilities established. Under these circumstances, a field kennel will have to be established.

Qualifications
Although teams do not need to be certified at home station prior to deploying, the kennel master/trainer must ensure each team is fully qualified in all required tasks, e.g., patrol disciplines, detection, and be subject of a current validation.

DEPLOYMENT

The QFEBP should be in place at the deployed location to establish the kennels prior to the arrival of any MWD teams. The UTCs required equipment should arrive with them at the AO to allow the kennel facilities to be operational in the minimum amount of time. If airflow does not allow for the full LOGDET to be shipped with the UTC, each MWD team must deploy with the following as a minimum:

Equipment for Deployment
1. Two (2) full sets of MWD gear.
2. Suitable MWD shipping crate or K-9 mobility container.
3. One (1) feed pan.
4. One (1) water bucket.
5. One (1) five-gallon water can.
6. 30-day supply of prescribed dog food per dog.

Security
Upon arrival at the AO, security is the first priority. The operations section will determine how the MWD teams will be integrated into the base defense.

Kennel Site
The initial kennel site should be determined during the leaders' recon or by a map recon prior to actual deployment. The actual site will most likely be determined when the kennel master arrives in the AO. When selecting a site the following guidelines must be considered:

1. The ground should be graded or have a natural slope to prevent standing water.
2. The area should be generally quiet to allow MWDs to rest when not working.
3. Adequate shade and ventilation must be provided. This can be provided by natural cover such as trees or constructed by using camouflage netting or tents. The use of fans can help with the ventilation of the kennel area.
4. An adequate potable water supply must be available at the kennels. Each MWD team will require a minimum of 10 gallons of water per day.
5. If a veterinarian is deployed, he must be consulted for any possible health hazards.

Construction of Kennels
There are numerous ways a field kennel can be designed when there are no permanent facilities available. The actual site and terrain as well as the number of dogs and materials on hand will be the determining

factors. The kennel area should be located in a relatively quiet area with minimal traffic to allow the MWDs to rest. If the kennels must be located in a congested area to ensure safety, a temporary screen or fence will be needed.

Shipping crates

Shipping crates can be used to construct a field kennel. The crates must have holes (approximately one inch diameter) on the top of the crate. The crate is placed upside down and raised 4 to 6 inches off the ground to allow drainage and to reduce parasite-breeding places. Place duct boards in the crate to prevent the MWD from injuring its feet or legs in the air holes. Small smooth gravel should be placed under and around the crate to allow for drainage and easy removal of solid waste. Place the crates under some type of cover, either natural such as trees or artificial such as a tent or camouflage netting to keep the MWD out of direct sunlight and provide some protection from inclement weather.

K-9 mobility containers

K-9 mobility containers are another type of kennel that may be used. Although these kennels are more modern, the same requirements for establishing the field kennel apply.

Satellite site

Based on the local threat, a satellite site may be needed to kennel MWDs at two or more locations to reduce the threat of losing all assets in one attack.

Support Facilities

Within the kennel area numerous support facilities and MWD specific areas must be established.

Bivouac area

When circumstances dictate, a bivouac area should be established within the kennel area but located separately from the actual kennels. At no time while under field conditions will MWDs and personnel be housed in the same tent.

Food preparation/utensil cleaning area
A food preparation/utensil cleaning area must be established. In this area rodent prevention measures must be taken to prevent the contamination of MWD food and feeding utensils. MWD food will not be stored in the same area with the MWDs.

Supplies and extra equipment area
An area must be established to maintain supplies and extra equipment.

Designated break area
A designated break area must be established. This area should be away from the kennels and must be kept clean of MWD waste at all times. A bucket or plastic bag can be used to store the waste until it can be disposed of properly. This will help prevent parasitic infestation within the kennel area.

Perimeter Security Measures
Security measures must be taken within and around the kennel area. A perimeter must be established around the area to prevent unauthorized personnel from entering. Rope or fence with signs stating "Keep Out" in English and host nation language will be used. When establishing the perimeter, remember that using items such as concertina wire to keep people out will also keep you in during an emergency. Determine if there may be a need for rapid egress when constructing the perimeter. If kennels are within a hostile fire/combatant zone, defense-fighting positions should be integrated into the overall kennel construction. Ensure adequate protective shielding is constructed to protect the MWD kenneling area from fragmentation debris. SF leadership must have in place procedures to protect MWDs in the event of chemical or biological attacks. Refer to USAF Manual 10-2602, Nuclear, Biological, and Conventional (NBCC) Defense Operations and Standards, dated May 03, Chapter 4, paragraphs A4.13 (Security Forces) through A4.13.6 for SF/MWD operations within a chemical/biological environment.

Standard Operating Procedures

The kennel master must establish SOPs for the kennel operations to run smoothly. The following are only the minimum areas that must be addressed:

Work/rest schedule

The mission will determine how the work/rest schedule will be designed. MWDs are fully capable of working 12-hour shifts with the understanding that they will need intermittent breaks. Alternating duties during a shift will also help keep the MWD fresh and alert. An example of this would be starting a shift at an ECP and after 2 hours rotate the team to a walking/mobile patrol in the cantonment area and then to a mobile reserve team later in the shift. Another example would be posting on a listening/observation post (LP/OP) and after a (suggested time/no more than 4 hours, etc.) period of time have a patrol with an assigned MWD team swap posts with the LP/OP team. Time must also be set aside to conduct training.

Frequency checks

Frequency of kennel checks and procedures for reporting incidents, accidents, or breeches of security in the kennel area must be established in the SOP.

Feeding of MWDs

Establish procedures for the feeding of MWDs at least two hours before and not sooner than two hours after working. This schedule must be available for all personnel conducting CQ duties to view. The SOP must also outline procedures for the removal and cleaning of the feed pans. In austere environments, MWDs will need fresh water more often than the standard 4 hours due to contamination by insects and other debris. Water buckets will be cleaned and disinfected as needed. Guidelines must be in place to ensure that MWDs are only given small amounts of water immediately following a hard workout, such as patrolling or conducting aggression training. As a minimum, clean pans and buckets with hot soapy water, rinse and air-dry.

Kennel sanitation

In field conditions it is **imperative** the entire kennel area be kept clean. This includes cleaning and disinfecting each kennel and removing all trash from the area. Employ rodent control procedures for both discarded and stored food areas. If veterinary support is deployed, they will be consulted concerning how areas and items will be cleaned and disinfected.

Safety

Establish safety standards applicable in the kennel area and around other personnel. Ensure these standards are strictly adhered to.

Stray animals

MWD teams will not be used to capture stray animals nor will stray animals be kept in or around the kennel area. Local procedures must be established covering the use of service weapons to defend an MWD team from stray or wild animals.

Operations group meetings

The kennel master or trainer will attend all operations group meetings to answer any questions pertaining to the capabilities, limitations, and employment of MWD teams.

Orientation Training

After arriving at the deployment site and ensuring site security has been established, the kennel master along with the trainer will conduct orientation training with all deployed teams. This training allows the deployed kennel master the opportunity to evaluate all deployed MWD teams. This training will give the kennel master the information needed to advise the ground defense force commander on the best employment of each team. Orientation training also gives the kennel master the opportunity to establish rapport, control, and discipline with deployed handlers. This training consists of field problems, obedience, gunfire, aggression, and detection problems. Once security has been established, the kennel master/trainer must validate all detector dogs on all available odors to ensure environmental conditions have not degraded the MWDs ability to detect all trained odors. This does not need to be a full validation, merely a test of the MWDs abilities to detect required odors under the deployed conditions.

Contingency Plan

The kennel master must establish procedures to disperse MWD teams in case of indirect fire. A plan must also be made in the case of natural disasters.

Re-supplies

The kennel master must coordinate with the S-4 to establish re-supply of food, water, and equipment. Do not wait until supplies are exhausted to initiate this process.

CAPABILITIES AND LIMITATIONS

Capabilities

An MWD team's detection and warning capabilities are a combined result of the dog's superior faculties of sight, sound, and smell, all of which far exceed those of a human.

Limitations

Terrain

Trees, bushes, heavy underbrush, thick woods, jungles, hills, ravines, and other terrain features can obscure an intruder's scent pattern. Obstructions and high winds often split and divert the scent pattern making it much more difficult for the dog to locate its source.

Smoke and dust

Smoke and dust are also limiting factors in detection because they reduce the MWD's ability to use its senses.

Wind, temperature, and humidity

Wind, temperature, and humidity can affect the scent pattern and the dog. High winds and low humidity quickly disperse the scent pattern while hot temperatures and high humidity will cause fatigue in the MWD. Rain and fog will also reduce the MWD's ability to use its senses.

Standard set time

There is no standard set time an MWD is capable of working; your operations tempo, type of mission, and the dog's fitness level all affect work/rest cycles. A physically fit dog with adequate rest and intermittent breaks should be able to work as long as needed.

Training

Training is the most important limitation that can be controlled. It starts at the home station in the pre-deployment phase. MWDs must be proficient in all required tasks prior to deploying. Once security has been established at the deployed location, training should be initiated. The type of training should be geared toward the intended mission of the MWD team. Failure to conduct this training will decrease the effectiveness of the team. Initial training also allows the kennel master to validate the teams' efficiency and in turn select the best team for the required duties.

EMPLOYMENT

Patrolling

MWD teams can be used on both mounted and dismounted patrols to detect enemy presence, avoid discovery, and locate enemy outposts.

Dismounted patrol

The MWD team must join the patrol in time to receive the warning order and participate in all phases of planning, preparation, and execution.

- The handler gives recommendations for employment of the team.
- The MWD team must participate in patrol rehearsals.
- The rehearsal allows the patrol members to become familiar with the MWD's temperament and the team's method of operation. It also allows the MWD to become familiar with the scents of the patrol members as well as the noises and motions of the patrol on the move.
- When taking cover, patrol members must avoid jumping close to the MWD team. When approaching the MWD team, always approach from the handler's front-right side because the MWD is normally on the handler's left. This should be covered in the handler's patrol brief.
- At least one patrol member must be designated as security for the MWD team.

1. Working the MWD requires the handler's full attention on the dog and does not allow the handler to scan the surrounding area for any threat.
2. Handling an MWD severely reduces the handler's ability to effectively use a weapon.
3. The proximity of the security person to the MWD team will be determined by the handler. The handler must ensure the security person does not distract the MWD.
4. The handler must ensure the security person is knowledgeable of the MWD team's responsibilities and able to perform his duties in the close proximity of a MWD.
5. Prior to departing on a patrol, handlers must brief all members on the MWD teams' capabilities, limitations, and safety issues. This briefing must include actions the patrol must take if the handler is injured, killed, or incapacitated.
6. Because wind is a key factor in the MWD's ability to detect, the team will normally be positioned on the point or flank depending on the wind direction. If the wind is coming from behind, the patrol the team should be placed in the position that is most advantageous to the patrol.
7. When speed is essential, the team should be placed in the rear to allow the patrol to proceed as quickly as need be without the MWD posing a threat to the members.
8. The length of the patrol and the weather conditions will determine the actual amount of equipment and supplies that will be needed. As a minimum, a muzzle, first aid kit, and extra water for the MWD must be taken. The muzzle will be used if the handler or MWD are injured to prevent the MWD from biting any of the patrol members. Patrol members may be needed to assist carrying the extra water for the MWD.

Mounted patrols

MWD teams may be attached to mounted patrols. When assigned these duties, the preparation for the patrol is the same as dismounted patrols. MWD teams should be assigned to a vehicle large enough to allow room for the handler to safely control the dog.

- MWD teams may be used to search vehicles along routes and at road-blocks for explosives.

- MWD teams can be used to provide security for convoy vehicles if attacked or for any reason there may be to dismount.

- After the all clear is given, MWD teams can be used to assist with search and clear operations after an attack.

- In some locations it may be beneficial to have an MWD on a mounted patrol to deter host nation personnel from approaching and reaching into the patrol vehicles.

Observation Post/Listening Post

MWD teams are most effective on these posts during the hours of darkness and times of limited visibility.

- The team must have a member assigned to act as a security person. Ample time should be spent with the security person prior to assuming post to associate him with the dog and to provide a patrol briefing.

- The team is located forward of the tactical area of operation to reduce distractions, preferably downwind of avenues of approach to allow the MWD to use its sense of smell. If the wind direction is not favorable, the MWD may still provide early warning using its sense of sight and hearing.

- The kennel master should be consulted when selecting sites for MWD teams. One or more alternate positions covering the same avenue of approach should be determined to allow the team to periodically change locations. This will maximize the MWD's tentativeness and reduce the chance of the MWD becoming complacent.

- As a minimum, each team posted on an LP/OP post will have a muzzle, first aid kit, and extra water for the MWD.

Support Capabilities

MWD teams can be used in a number of ways within the AO both as a psychological and physical deterrent. Using MWD teams at different locations for short periods of time can give the perception that MWDs can be anywhere. The fear of dogs alone may prove to be an effective deterrent.

- MWD teams offer both a physical and psychological deterrent in EPW and detainee operations. MWD teams can be used at collection points, holding areas, during movement, and enhancing perimeter security at a compound or camp. MWDs can help locate and capture escaped EPWs. Under no circumstance will MWDs be used in the interrogation or interview of EPWs or detainees.
- MWD teams can be used to patrol, both mounted and dismounted, assigned areas within the cantonment area to conduct security checks, respond and clear unsecured buildings, and assist at roadblocks controlling personnel and detecting contraband.
- MWD teams placed at entry control points can search all incoming vehicles and cargo for contraband as well as provide a psychological deterrent.
- When assigned to a response force element, MWD teams can provide a quick response. In this capacity they can be used to investigate alarm, sensor, and tripwire activations as well as possible sightings in unauthorized areas.
- When using MWDs in confrontation management situations, the presence of an MWD may provide a psychological deterrent, however, the mere sight of an MWD may escalate the situation.

1. MWDs will not normally be used for direct confrontation with demonstrators.
2. During peaceful stages, MWDs should be held in reserve and out of sight of demonstrators.
3. If the situation deteriorates, MWD teams should be moved up within sight of the crowd, but still at a distance.
4. Only when actual physical confrontation erupts should using an MWD team be considered. When a team is committed, all other personnel should be positioned at least 10 feet from the MWD team due to the dog not being able to distinguish between friend or foe.
5. MWDs should never be released into a crowd.

RE-DEPLOYMENT

The same considerations apply as when making pre-deployment travel arrangements. Ensure there are no animal restrictions or quarantines at any enroute stops.

Ports of Entry Inspections

MWDs are subject to inspection at ports of entry and may be denied entry into the United States if they have evidence of an infectious disease that can be transmitted to humans. If a dog appears to be ill, further examination by a licensed veterinarian at handler's expense might be required at the port of entry. MWDs must have a certificate showing they have been vaccinated against rabies at least 30 days prior to entry into the United States as well as a health certificate. There are no restrictions on who (nationality) can sign a health certificate or rabies vaccination certificate, provided the individual is licensed to practice veterinarian medicine in that country. Dog's medical records must also contain evidence of a favorable FAVN test.

Adequate Supplies

Each handler must ensure he has an adequate supply of dog food in case of any unforeseen stops enroute to the home station. In addition, handlers must hand carry any MWD medication, MWD's medical records, and at least one set of dog gear.